Construction equities: evaluation and trading

Construction equities: evaluation and trading

Fred Wellings

International Equities Series

WOODHEAD PUBLISHING LIMITED

Cambridge England

Published by Woodhead Publishing Limited, Abington Hall, Abington, Cambridge, CB1 6AH, England

First published 1994, Woodhead Publishing Limited

© Woodhead Publishing Ltd

British Library Cataloguing in Publication Data
A catalogue record for this book is available from the British Library

ISBN 1 85573 109 6

Designed by Andrew Jones (text) and Chris Feely (jacket)
Typeset by Best-set Typesetter Ltd., Hong Kong
Printed by St Edmundsbury Press, Suffolk, England

Contents

Acknowledgements

Many people were kind enough to read through, and comment on, the draft either in whole or part, and supply me with additional statistical material. I would particularly like to thank Barry Bisset, Duncan Brand, George Capon (to whom I owe a particular debt on the accounting sections), Peter Cardell, Adrian Coles, Martin Davis, Tony Good, David Green, David Holliday, Gary Marsh, Michael Rubie, John Stewart, Rory Sweetman, Sarah Sweetman and Jeremy Withers Green. Especial thanks are due to Rosemary Ackland and Simon West of Credit Lyonnais Laing's library and to my neighbour Mary King who typed most of the manuscript for me. As always, the mistakes and opinions remain mine.

*In*troduction

The construction sector, depressed as it has been by its worst recession since the war, represents only 1% of the Financial Times-Actuaries Index by market capitalisation. Yet the industry has an importance to both the economy and the stock market which extends far beyond that modest percentage. In 1990, its peak year, the output of the industry accounted for 6.8% of Gross Domestic Product and, including housing, just over half of gross domestic fixed capital formation. Even in 1991, one and a half million people were directly employed in the industry out of a national total of 27 million.

The industry's influence reaches through the housing market into every household and, at the other end of the economic spectrum, to the nation's infrastructure. Apart from offering secondary employment for the building materials industry, builders' merchants and d.i.y. outlets, construction provides the stimulus for derived demand throughout the consumer goods markets: white goods, carpets, furniture, etc. The fortunes of the construction industry have therefore much wider import than suggested solely by the range of medium-sized companies which comprise the FT-A contracting and construction sector.

These actuarial classifications are, of course, narrowly defined. Many

companies outside the sector have significant interests in both contracting and housebuilding, for instance BICC (Balfour Beatty), Hanson
(Beazer Homes), P & O (Bovis), Tarmac (Tarmac Construction and
McLean Homes), and Trafalgar House (Ideal Homes, Trollope & Colls
and Cementation). Although not covered as a separate industry in this
book, the building materials sector is directly governed by the economics
of the construction industry, which is discussed in the body of the text,
and the building materials sub-sector accounts for some two and a half
per cent of the FT-A Index. It should also be remembered that the
boundaries of process engineering are somewhat ill-defined and such
activities that exist happily within construction companies are also
found within the engineering sector, e.g. Babcock International, Haden
MacLelan, Simon Engineering and Whessoe. Indeed, the chapters on
contract structure and accounting are as relevant to the engineering
industry as to the construction industry.

This book is targeted at those who are approaching the analysis of the
construction industry for the first time. They may be employed within
the investment community and have been given the construction industry as a new specialisation; alternatively, they may work in industrial
market research, either directly within the construction industry or in
one of the multitude of enterprises for whom construction still represents
an important market. There is such a body of literature on the construction industry that a large part of the book is a distillation of what is
available elsewhere; its virtues lie in summary and the grouping of the
various disciplines under one set of covers. There are some topics less
well covered elsewhere, such as housebuilding accounting, and there are
specialist sections in the book which will also be of help to those who
already possess a working knowledge of the industry.

The history of the construction industry, which provides the book's
opening chapter, is perhaps the one topic which the time sensitive
construction analyst may ignore, although that would be to the detri-
ment of any long term perspective of the industry; anyway, the author
enjoyed writing it. Moving on from that is a discussion of the role of the
contractor as a developer: what is the rationale for the contractor being a
property developer or housebuilder? In contrast to other areas in the
book, it is not a topic on which one sees much written and the author
has long regarded it as an area of confused thinking, both within and
without the industry.

The next substantial chapter is on construction statistics and their
sources. There are at least two 400 page and one 600 page books on this

specific topic (see Bibliography), so this chapter cannot claim to be exhaustive. Instead, this book pays particular attention to the high profile statistics, especially the Department of the Environment's figures for construction output and orders, and housing starts and completions; other housing series included relate to gross housing sales. The emphasis throughout is on how to interpret those statistics analytically, when not to take them too literally and, on occasion, which competing alternative to select. To the extent that interpretation rather than description is involved, there is an overlap with the material contained within the chapter on forecasting.

Two statistical areas are developed into self-contained chapters. The first covers house price indices and house price earnings ratios. These are an important part of any analysis of the housing market and scarcely a week goes by without one series being quoted in the media. Unfortunately, there is a wide variety of choice of house price data, not always giving the same answers and certainly not accurate to the first decimal place of percentage change that is typically quoted. House price earnings ratios are subject to even more error in that they include the earnings measurement as well as the price, and there are serious theoretical problems in deciding whose earnings should be used in the calculations. The chapter seeks to outline the leading series, the bases of calculation, their comparability and their limitations.

The other set of statistical data to be treated to a whole chapter is that relating to the long term demand for housing and in particular the population and household formation statistics. The chapter analyses the sources of the data and the extent to which they can be used as the basis of a long term forecast for the housing market. The author points out that whereas the official medium term population forecasts have a reasonable degree of accuracy, those for household formation have proved distinctly wide of the mark; he is far from convinced that they can provide anything other than a broad brush overview of the long term requirements of the housing market. Nevertheless, many reports on the long term housing market treat the household formation forecasts as definitive and, however sceptical the analyst, it is important to be well versed in their pedigree.

Reverting back to the contractors themselves, the next chapter summarises the types of contract that the analyst will encounter, and attempts to distinguish between those elements of the contract that are based on what the contractor does – 'the involvement criteria' – and those based on how he is remunerated – 'the payment criteria'. Other contractual matters covered are those which may have a material bearing on profitability, including variations and claims, liquidated damages and bonding.

Naturally following on from the contract structure is the chapter on construction accounting. This assumes a general knowledge of accounting issues and deals primarily with those matters which are peculiar to the construction industry, particularly the valuation of long term work in progress. A substantial element of the chapter is actually devoted to housebuilding accounting, partly because there is less available in the general accounting books, and topics such as the accounting treatment of incentives on sale, capitalised interest, land provisions and options are discussed.

The chapters on forecasting, covering the industry and the companies, has caused the author the most problems despite, or perhaps because, that is how he has earned his living. Most economic and industrial forecasters will honestly admit in conversation that forecasting is an extremely uncertain process and that in the real world the best one can do is to make sure that one's guesses are based on a firm grasp of the relevant statistical sources and a sound understanding of the issues; at least the professional forecaster knows where he went wrong. There are, of course, techniques that one can use, typically based on past or presumed relationships between the item to be forecast (say, housing starts) and either a published lead indicator or another statistic for which the forecast is taken as given (interest rates). A wide selection of these industry relationships are discussed in the book, although few are found to work consistently in each cycle.

There is also a chapter which discusses the even more conjectural world of profits forecasting, distinguishing between profits earned and profits published in companies' statutory accounts, and the outlook for contractors' cash flow. Readers of this chapter may not become better forecasters but at least they should be able to argue their case more cogently, and be better placed to say 'that does not necessarily follow' when presented with other people's forecasts.

Finally, there is a chapter devoted to the companies concerned, listing the contractors by turnover and the housebuilders by unit sales. The contracting sub-sector of the FT-A Index is compared with interest rates, as one example of the relationship between a supposed leading indicator and share price movements.

The development of the modern construction industry

THE CHANGING WORKLOAD

The much-used expression 'food and shelter' is testament to the essential priorities which are satisfied by the construction industry. In both primitive and advanced economies, construction is one of the basic industries; it owes its existence to man's need for shelter not only for himself but also for the multitudinous activities in which he engages. It existed to support his agriculture through irrigation projects (one of the earliest engineering problems derived from the flow and measurement of water), milling, and farm buildings. It supplied his need for communication from Roman road building, harbours, through the eighteenth century canal and nineteenth century railway building manias to twentieth century airfield construction. Some of the most enduring structures survive as a testament to religion – pyramids, temples and cathedrals – and to war – Iron Age forts, Hadrian's Wall, and medieval castles.

As society changes and develops, so also does the composition of the construction work and the nature of the client. The great civil engineering workloads of the late eighteenth and nineteenth centuries, canals, railways and docks, are either no more or but a shadow of their earlier selves. In their place have come nuclear power stations, motorways and airport terminals. The water and sewerage systems built by the Victorians have returned as a fashionable investment area. Industrial investment, once centred on textiles, iron and steel, now serves the manufacturers of computers, telecommunications and pharmaceuticals. Hotels followed churches; out-of-town retail 'sheds' replaced department stores and, at its peak in 1989, office building was accounting for 19% of the industry's total new orders.

THE CHANGING CLIENTELE

The economic status of the client has also changed. Nowhere is this more evident than in the housing market where, before the First War, the private landlord accounted for 90% of the housing stock. The inter-war period saw the first great thrust of owner occupation, but this was followed after the second war by an unprecedented programme of local authority housing construction; the public sector accounted for a consistent 86–87% of housing completions between 1948 and 1952. Since the mid-1970s, local authority housing programmes have been progressively reduced to the point where the sector exists as little more than a memory in the statisticians' working papers.

The election of the first majority Labour government in 1945 also saw the State become the dominant customer for the civil engineering component of the construction industry as transport infrastructure spending was increased and basic industries were brought under nationalised control, to the point where the construction industry was widely regarded as the natural vehicle for successive governments' regulation of the economy. Since the early 1980s there has been a dramatic reversal of the State's role in construction as industry after industry has been privatised (see Table 1.1).

A recognised by-product of a mature economy is that the proportion of national income spent on infrastructure gradually declines but as the absolute size of the capital stock continues to grow, so repair and maintenance assumes a more significant role (see Table 1.2).

Table 1.1 *Change in composition of new work, %*

Total new work	1935	1955	1971	1981	1991
Housing					
Public	6	30	16	10	3
Private	36	21	21	20	15
Total	42	51	37	30	18
Non-housing					
Public		23	33	29	24
Private		26	30	41	58
Total	58	49	63	70	82
All Public		53	49	39	27
All Private		47	51	61	73

Source: DoE from 1955. Figures for 1935 are taken from Bowen & Ellis, 'The Building and Contracting Industry', Oxford Economic Papers, No. 7 (1945). Their output analysis is based on the 1935 Census of Production.

Table 1.2 *Growth of repair and maintenance*

	1935	1955	1971	1981	1991
New work	80	62	72	64	58
Repair and maintenance	20*	32	28	36	42

* Building repair and maintenance only.

EVOLUTION OF THE TWENTIETH CENTURY CONTRACTOR

The organisation of the construction process has also seen profound change as the independent tradesmen made the transition to the modern corporate builder and the work of the great engineers became the province of the civil engineering contractor. It is at this point that a distinction should be drawn between building and civil engineering though, like all trades which overlap, the distinction is easier to recognise from the use of examples than from a clinical definition. However, building can be thought of as the creation of structures from existing components, and is normally above ground e.g., houses, offices, shops, etc, and will frequently involve elements of repetition. Civil engineering (as it eventually became known to distinguish it from the earlier military engineering) encompasses the large scale movement of materials where, although the principles may be standardised, the operations are invariably of an individual nature. Most ground and below ground work is regarded as civil engineering, examples being roads, railways, reservoirs, foundations for buildings, harbours, tunnelling, pipelaying and so on. The professional supervision of these two branches of the construction industry are distinct, building being the province of the architect, and civil engineering of the consulting engineer. The contractor also tends to keep the building and civil engineering functions as separate entities, a distinction which can be observed throughout the eighteenth and nineteenth centuries as the modern contractor evolved.

Dating the evolution of the modern builder can only be approximate but J.H. Clapham, in his seminal *Economic history of modern Britain*, identified a change in the organisation of the building industry from the late eighteenth century and, indeed, even earlier in London. 'Right through the nineteenth century, in the full flood of capitalism, there was no industry in which the handicraftsman more frequently rose to be the small jobbing employer and perhaps, eventually, a builder on a large scale.' (*The early Railway Age*, p. 162)

In his history of the building trades unions, *The builders' history*, R.W. Postgate describes the transition from craftsmen working on their own account to working in groups and it was this massing of the specialist labour force that was the essential prerequisite to the development of the general builder.

> *It is not until the eighteenth century that we find anything more than ephemeral combinations of journeymen, and in the building trades there are indeed very few traces of them before 1800. It was in the eighteenth century that the capitalist system . . . spread all over England and Scotland. The century was marked by the complete disappearance of the traces of the medieval guild system, and the appearance of large establishments in all trades in which one master directed far more journeymen than could ever hope to become masters in their turn.*

The builder

In building, as opposed to civil engineering, the work continued to be executed by the specialist trades – bricklayers, carpenters, painters and so on – but it became increasingly common for one of the trades, typically the carpenter, to take the building contract and let out work to the other trades. Clapham dated the emergence of the term 'builder', in the sense of a commercial organiser of the construction process, to the second half of the eighteenth century.

Tension between the builder in his capacity as principal contractor to the client, and the tradesmen or sub-contractors is nothing new. Postgate describes the strains which the emergence of the builder from amongst the ranks of his fellow trades had caused. In 1833 the Great Operative Builders' Union made an attack on 'the new system of general contracting'. 'The master builder – sometimes calling himself an architect or a master-carpenter – who tended a general estimate for the erection of a large building . . . was well hated by the small jobbing master.' Resistance was greatest in the North: in Manchester master masons demanded 'that no new building should be erected by contract with one person'.

The first half of the nineteenth century clearly saw a process of organisational change during which the modern structure of the building process was founded. Clapham:

> *By 1830 'the respectable builders' of London were already specialised into definite groups. There was a small group . . . who did little but erect public buildings. A second, larger, group devoted themselves to the building of shops and business premises. A third, perhaps not all respectable, came . . . those who took risks with private houses, 'speculative builders' as they were already called. They were no new type. (The early Railway Age, p. 164)*

Thomas Cubitt (1788–1855) is generally credited with developing the first large scale building organisation which could carry out the totality of a complex contract. In his review of the industry in London at the end of that period, *The London building world of the 1860's,* John Summerson refers to substantial building works being done by general contractors in the London area and encompassing all the necessary trades. Describing the sizes of contractors he says:

> *In the top layer only . . . were general contractors in the sense which Thomas Cubitt had pioneered as long ago as the 1820's. That is to say, businesses which incorporated all the trades, each trade conducted by its own foreman. In principle there was no sub-letting, a practice frowned upon by architects as leading to irresponsible cut-price workmanship.*

9

A more detailed description of the emergence of the general building contractor is presented by E.W. Cooney in 'The origins of the Victorian master builders' (*The Economic History Review 1959*, vol. 8, pp. 167–176). Indeed, Cooney finds evidence of large building enterprises predating that of Cubitt, instancing men such as Alexander Copland building barracks in the late eighteenth century and employing 700 men. But, whoever takes the credit, he concludes that 'In the space of half a century an industry which had been organised primarily on a craft basis had, without the stimulus of any important technological advances, thrown up a group of large, complex, and markedly capitalist businesses.'

Cooney's explanation for the development of such building firms rests firstly on the increase in demand and secondly, on changes in the form of agreement between client and builder. Cooney examines the idea that 'During the first half of the nineteenth century, the industry's customers, including public bodies, came to believe that the best basis on which to arrange for building was to obtain competitive tenders for work to be carried out by one builder at a fixed cost.' Cooney argues that the larger builders with substantial capital were better positioned than the smaller master craftsmen to meet the greater risks and outlay involved. This type of contract was not common in building before the nineteenth century: 'There was, in fact, general agreement among seventeenth and eighteenth century architectural writers that to put all the work into the hands of a single "undertaker" was to court disaster.' (Quoted from H.M. Colvin, *A biographical dictionary of English architects 1660–1840*, published 1954)

The building firms that emerged during this period were by no means insubstantial, though they have left a disappointingly small amount behind in the way of corporate history. Cubitt and Copland, mentioned earlier, both employed around 700 men on general contracting although Thomas Cubitt later concentrated his resources on the London developments for which he is now best known. Perhaps the largest builder by the middle of the century was George Myers (1803–1875) who started as a stonemason but became a leading builder of churches, cathedrals, hospitals and large houses. On one contract abroad, building Ferrieres (near Paris) for Baron James de Rothschild in the 1850s, he took a labour force of 400 men with him. The Ferrieres contract was an uncommon occurrence for a builder; transport restricted most builders to their immediate localities, hence the London builders were better recognised, but there were also regional firms of significance who were capable of working at distance for prestigious contracts, e.g. the Bristol firm of James Diment (later Stephens Bastow) who worked as far north as Warwickshire and across into London.

The civil engineer

Although it has been indicated that reasonably sized building firms could be found in the nineteenth century, in practice the builders were predominantly small and localised and there is little evidence of them working abroad. The engineers, in contrast, operated on a far grander scale. Ask a layman to conjure up the name of a nineteenth century builder and you might get Thomas Cubitt but no more. Ask the same question about an engineer then the great names trip off the tongue – Brassey, Brindley, Brunel, McAdam, Smeaton, Stephenson, and Telford. Many straddled the worlds of mechanical engineering and civil engineering in a manner which could not now be contemplated by an individual nor even found in very few corporate enterprises. Unlike the builder, the civil engineer was equally at home in the furthest reaches of the Empire as in Britain and, indeed, there were men like George Pauling and Sir John Norton-Griffiths who did virtually no domestic work yet were major contributors to the construction of the African railways.

The civil engineering industry operated on an altogether grander scale than did building. The composition of its workload probably changed more radically as the generations passed:

> *Engineering before about 1760 had been confined mainly to naval and military demands by the Government – the constructional requirements of our early dockyards and harbours, of land fortifications and of roads built for the better travel of troops rather than for civil needs.* (Norrie, *Bridging the years – A short history of British civil engineering*)

Subsequent concentrations of activity were to be found in the turnpike roads (built by men like McAdam and Telford) in the late eighteenth century, the canal mania of the 1790s and the railway mania of the 1840s, periods in which the workload of the typical engineer would have been almost entirely dominated by one class of client.

Amongst the duties of the early civil engineer, that of design appeared foremost. Construction was carried out by the direct administration of services and labour and the 'engineer' could be both the manager and the agent for the promoter. In Smeaton's era:

> *The day of the general contractor ready to assume constructional and financial responsibilities had not then arrived. Contract workers consisted mostly of labour gangs who toured the canal making areas and undertook excavations and other works at piecework rates. These men ... were eventually to become the chief resource of the industry.* (Norrie, p. 23)

From these came the individuals with the character and ability to direct large operations and able to attract financial backing. 'Such men started the constructional industry as we know it today and the contract system evolved out of the inconsistencies as much from the rigidities which its own legalities created.' (Norrie, p. 23).

The nineteenth century saw the evolution of the civil engineering industry from the casual labour of the canal era:

> *promoters of public works ... demanded binding commitments to cover all the risks and responsibilities of construction from those who were to carry out their projects and sought proper guarantees against a contractor's inability or refusal to fulfil his obligations ... they tended to make a contractor something more than a supplier and organiser of labour. The equitable conditions of contract which were introduced under Telford's influence, the readiness of the merchant bankers of the period to support this new form of risk enterprise and the entrance of men of great calibre to carry the executive burden, all helped to raise the status of public works contracting.* (Norrie, p. 89)

Just as there is a distinction between the architect who designs and supervises, and the builder who organises and constructs, there exists a similar relationship between the engineer and the civil engineer. However, the distinction between the engineer as the designer/supervisor and the engineer as organiser of the construction process was far from clear for much of the nineteenth century. Of the engineers' names mentioned earlier, some designed and supervised, some were contractors, and some were both. Indeed, individuals readily switched their roles between contracts, perhaps being the engineer here or the contractor there. Perhaps the greatest confusion of identity was created by the railway building boom which, in turn, did much to stimulate the development of the civil engineer as an independent contractor.

The profusion of railway schemes stretched the organisational resources of the promoters and their engineers:

> *they had every interest in the rise of the new men, the contractors, whose ability was organisation and command, who could fix their price for a length*

12

> *of line, find their workmen, plant and materials, build to the engineers'*
> *specifications, and deliver the finished work on time at an agreed date. Such*
> *men were rare.* (Middlemass, p. 34)

The contractor's main task was to keep the men employed and the railway age did much to encourage the contractor as organiser rather than engineer.

> *With regard to the large contractors, they were not unknown before the*
> *railway era. The construction of inland navigation, and of docks and*
> *harbours, had called them into existence to a limited extent; and Messrs.*
> *Joliffe and Banks, for instance, were large men of that class. But with the*
> *commencement of the railway system began an age of great works, during*
> *which undertakings of far more colossal dimensions were rapidly projected,*
> *and required to be as rapidly carried into execution.'* (Letter to Arthur
> Helps from John Hawkshaw quoted in *Life and labours of Mr.*
> *Brassey*, 1871)

While the civil engineering contractor may have been establishing himself as an independent organisational entity, he was nevertheless frequently financially embroiled in the projects he served; indeed, he may also have played a key role in sponsoring and financing those same projects – a stance well familiar to those involved in the contemporary EuroTunnel – and with no less embarrassment. Promoters in the mid-nineteenth century frequently obtained contractors' support in return for shares in the venture and, of course, the contract. This was criticised in Britain but abroad 'the financing of a project was often an accepted form of risk enterprise under concessions which ensured that some check on capital and operational expenditures was achieved.' (Norrie, p. 99).

The comparative size of the engineering as opposed to the building contractor was touched on earlier. The largest builders perhaps approached 1000 men under employ but single engineering contracts could dwarf that figure: Robert Stephenson had up to 20 000 men working for him on the London and Birmingham Railway. Thomas Brassey (1805–1870), the largest civil engineering contractor in the mid-nineteenth century employed nearly 100 000 men at his peak. Peto & Betts was employing 30 000 when it collapsed in the wake of the Overend & Gurney Bank failure in 1866. Numbers such as Brassey employed have probably never been exceeded since in a single civil engineering organisation as mechanisation gradually replaced manual tasks; Weetman Pearson, reputed to be the largest contractor in the world at the start of the twentieth century, at his peak employed only around half Brassey's numbers.

The construction industry gradually moved towards a more recognisably modern format in the last quarter of the nineteenth century. Contract risks lessened as the structural sciences developed. Tendering methods became more standardised, safeguarding the contractors against bigger risks, but at the cost of lower profit margins. The architect assumed a more important, and independent, role and the development of local government at the end of the nineteenth century brought new procedures for awarding contracts.

The contractors themselves had moved on from the organisation of labour and traditional crafts:

> *there had also come about what amounted to a revolution in the composition and the extent of the assets in a contractor's organisation. From about 1850 onwards there had been an ever increasing use of steam powered labour saving and heavy plant . . . This had entailed a larger employment of experienced engineers and other skilled services in progressive firms.* (Norrie, p. 100)

Men were no longer automatically dismissed at the end of a contract, but formed a permanent labour force:

> *When he first came into the business [1876], it had been the habit of contractors to wind up the services of the entire staff at the end of a contract, and to engage new staff for each contract as it came along. Pearson from the beginning decided to retain the services of men who had proved themselves, and brought them into the office and paid them salaries until the next job was ready for them.* (J.A. Spender, *Weetman Pearson First Viscount Cowdray*, 1930)

A second generation of contractors appeared, men like Sir John Aird, Weetman Pearson, Sir John Jackson and Sir William Arrol.

Victorian survivors

A surprising number of Victorian contractors survive to this day. At the time of the National Federation of Building Trades Employers' Centenary in 1978, some 285 of the 11 000 members had been in existence 100 years earlier, albeit not in the same form. From today's national contractors, these included John Laing, John Mowlem, Trollope & Colls (now in Trafalgar House), Wates and, much altered, Holland Hannen and Cubitts (now Tarmac). However, most of the nineteenth century contractors did not long survive their proprietors. George Myers sold his business to his sons for an annuity and little more was heard of it. After Robert, the Stephensons had no succession, no more than did such other 'engineers' as Telford and Rennie; the Gibb and Rendell families turned themselves into consulting engineers. The railway contractors Brassey and Firbank left only their tracks behind them. More dramatically, Peto went under in 1866 with liabilities of £4 million, Sir John Aird's sons took only two years after their father's death in 1911 to accept a ruinous dock contract in Singapore and the firm was liquidated, and Sir John Norton-Griffiths shot himself on the Nile in 1930 as his 1929 tender for the Aswan Dam heightening threatened him with bankruptcy. But undoubtedly the most clear cut view on contracting succession was expressed by Weetman Pearson. Pearson took on his first contract in his grandfather's small firm in 1876 and went on to create perhaps Britain's leading international contractor by the First World War. Before his death in 1927 he deliberately closed down the contracting part of the business: 'Contracting, as he used to insist, was a one-man job . . . he was clear that the brilliant and undefeated chapter of S. Pearson & Son, Contractors, must be closed on his departure.' (J.A. Spender, *Weetman Pearson First Viscount Cowdray*.) There have been many contractors in more recent times who have found the succession more complicated.

After the First World War those of the nineteenth century contractors who were still active were predominantly corporate in structure and there was little overlap with the nineteenth century promotional function, and the professional roles of the consulting engineer and architect stood quite separate. The contractor had a range of contracts at any one time which gave him a spread of risk and was far removed from the limited specialisations of fifty years earlier. Indeed, the type of work being done had also changed in emphasis. No longer were there the canal, turnpike or railway building booms; in their stead came the inter-war housebuilding boom, largely privately driven, and reaching levels of output never seen before. It was the inter-war housing boom which provided the growth for such firms as Taylor Woodrow and Wimpey: by 1936 Taylor Woodrow was building at the rate of 2500 houses a year (Alan Jenkins, *On Site 1921–71*, 1971) while Wimpey was building around

1000 units a year before the Second World War (private conversation with Wimpey director).

Rearmament

From 1936, rearmament began to move the balance of activity back towards civil engineering and the onset of war itself limited housebuilding for almost a decade. Wartime demands left builders with no alternative but to become civil engineers and the workload was dominated by airfield construction, camps, civil defence and such projects as the Mulberry Habour (for D-day).

> *The construction industry was transformed during the course of the war . . . rearmament and the war induced a series of large construction programmes and large projects, from which the existing large contractors considerably expanded. Former housebuilders . . . emerged at the end of the war as very large contractors.'* (Hedley Smyth, *Property companies and the construction industry in Britain*, Cambridge, 1985)

The post-war priority was for war-damage repair, reconstruction, and the promotion of public infrastructure under the country's first majority Labour Government. Once important inter-war housebuilders were concentrating on their general construction work and, as the first revival in new housebuilding activity came from local authorities, such housebuilding as they did was now viewed as an extension of their contracting rather than a development activity in its own right. Private housing only began to resume in earnest in 1953, further stimulated in 1954 by the ending of building controls. As often as not, the firms that took most advantage of the opportunities were post-war entrepreneurs with no previous corporate history, e.g. Northern Developments, Whelmar and, later, Barratt.

Forty years of change

From the mid-1950s, the output of the construction industry has been statistically documented in its present form, offering a continuity of presentation that stretches back nearly forty years. In the context of the

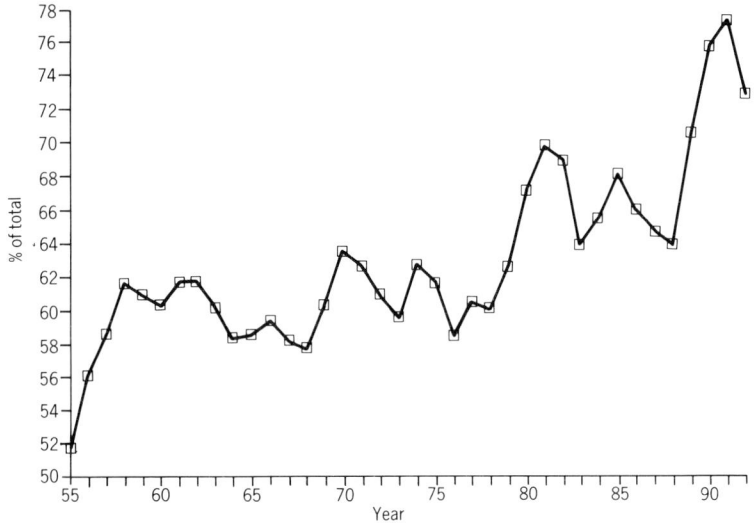

1.1 *Non-housing as a percentage of new work.*

history of the construction industry that may not seem long, but it does serve to illustrate the considerable changes in the workload that can occur within only one generation. As can be seen from Fig. 1.1, the importance of new housing has declined and, from just over half of total new work in 1955, non-housing construction accounted for 78% at its peak in 1991.

Within the non-housing sector, the industry has seen a remarkable rise in the importance of commercial property construction. The 1947 Town and Country Planning Act effectively nationalised development values, putting speculative development on hold for six or seven years until the abolition of the development charge in the Town and Country Planning Act of 1953 and of building controls in 1954.

> *The almost simultaneous breakthrough on two important fronts immediately freed the pent-up talents and genius of a small handful of men of vision and foresight . . . they had seen the enormous potential for redevelopment left by the war; they had watched the demand for good property in key positions steadily growing greater and greater. They had witnessed – and recognised what it meant for property – the evolution of a trend that amounted to a second industrial revolution: the advent of automation, and with it the rise of an*

17

> *enormous white-collar brigade wanting better office accommodation at work*
> *and more comfortable living conditions at home.* (Brian Whitehouse,
> *Partners in Property*, p. 21)

The first full year after the abolition of building controls saw commercial development at under 20% of non-housing output. By 1990 the
figure was almost 50% (see Fig. 1.2), and even that understates the true
impact of the property boom; an element of industrial building would
have been development led, while local authorities were often partners in
town centre redevelopment. The construction sector has had a near
incestuous relationship with the commercial construction sector and the
role of the contractor as developer is discussed later.

The housing market has also seen a profound change in its composition, as between the public and the private sector (see Fig. 1.3),
particularly since the late 1970s when the local authority housebuilding
programme was progressively reduced to little more than nominal levels.

Equally interesting has been the change of emphasis between new
work and repair and maintenance. The immediate post-war period of
reconstruction inevitably produced a high repairs content for the construction industry but, as the backlog of repair was reduced, and new
building freed of control, repair and maintenance fell to under 30% of
total construction output by the late 1950s and held at that level until

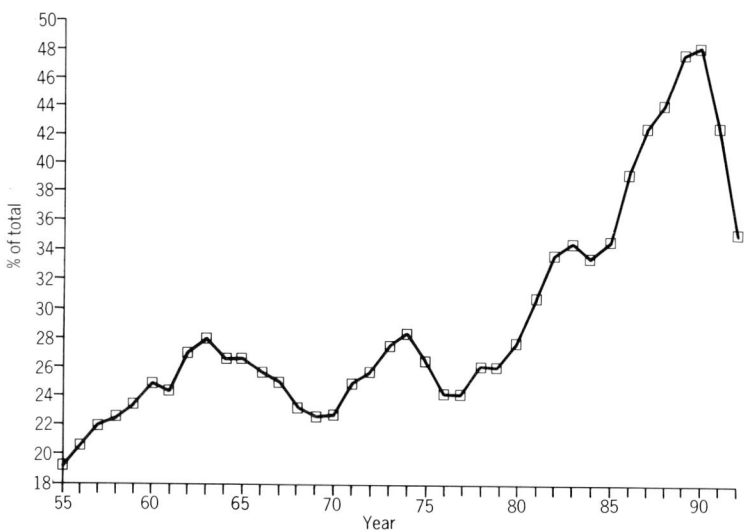

1.2 *Commercial output as a percentage of total non-housing new work.*

18

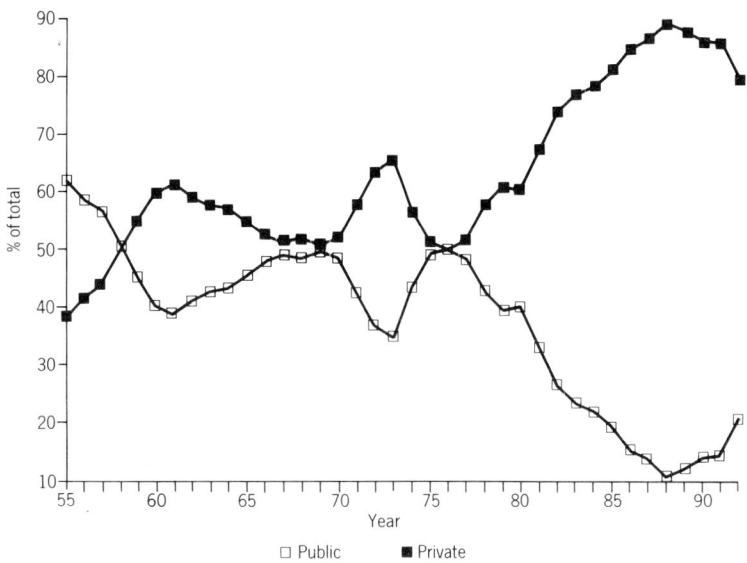

1.3 *New housing: public sector* vs. *private sector housing.*

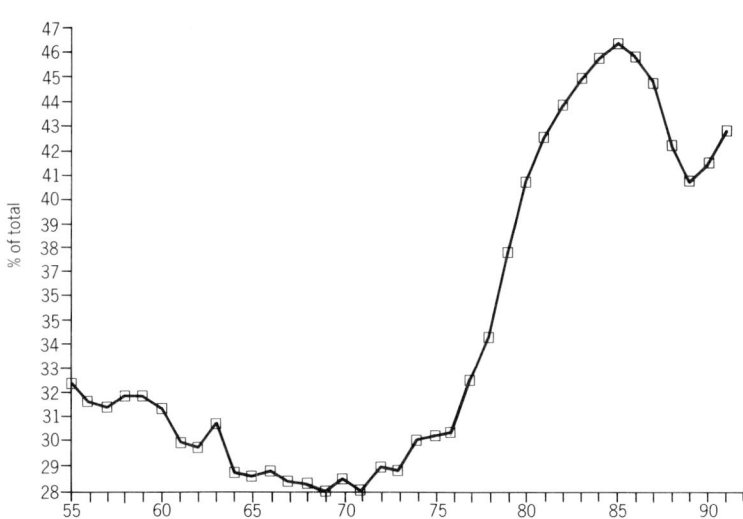

1.4 *Repair and maintenance as a percentage of all new work.*

the early 1970s. From that point, a secular decline in new construction and increasing concentration on improving the quality of the housing stock, drove up the proportion of repair and maintenance to a peak of 46% in 1985. That percentage only dipped in the late 1980s because of the commercial building boom and, with the collapse of the latter, so the repair and maintenance figure is creeping up again (see Fig. 1.4). Indeed, if allowance is made for unrecorded work (unregistered contractors and do-it-yourself) then repair and maintenance must now account for comfortably over half the industry's workload.

*T*he present structure of the construction industry

The *Annual Housing and Construction Statistics* contains a wealth of data on the number of firms by size, trade and region, the employment within those categories, and the value of work done. Enthusiasts for such data can refer to Sections 2 and 3 of the above. A bare minimum of statistics have been extracted to show what most people already know: that there is an enormous number of contracting firms, typically of one, two or three men, and organised by trade. Given the tradesman's well known predilection for filling forms, not all small firms may be recorded while, at the other end of the spectrum, some large firms may report parts of their organisation separately rather than as a consolidated unit. In addition to private contractors, public authority direct labour output and employees are also recorded. Particular care should be taken with trends, which can be affected by the number of firms included on the statistical register; for instance, the use of the VAT register in 1982 and 1983 disclosed a number of firms not previously included on the DoE's construction register.

EMPLOYMENT

The average number of people in employment in 1990 was 1.7 million; this excludes the building materials and merchanting industries.

There has been little overall change in the numbers employed in the industry (see Table 2.1) over the last twenty years although numbers have obviously fallen during 1991 and 1992. Within the total, the self-employed numbers have risen during the 1980s, probably reflecting trading opportunities in private housing and repair and maintenance, together with administrative and financial advantage. In contrast, the public sector direct labour halved over that same period as more work was put out to private competition.

Table 2.1 *Numbers employed in construction, 1991*

		000		*%*
Self-employed		648		42.5
Contractors				
Operatives	464		30.4	
Staff	231		15.1	
		695		45.5
Public authorities				
Operatives	120		7.9	
Staff	63		4.1	
		183		12.0
		1526		100.0

Source: Annual Housing and Construction Statistics, *Table 2.1.*

Table 2.2 *Numbers employed in construction, 1991*

Size of firm	*New work,* %	*Repair & maintenance,* %	*All work,* %
1–7	14.1	48.1	26.5
8–114	29.3	32.2	30.3
115–299	12.7	6.7	10.6
300–1199	22.6	9.5	17.8
1200+	21.3	3.5	14.8
	100.0	100.0	100.0

Source: Annual Housing and Construction Statistics, *Table 3.8.*

SIZE OF FIRM

With over 200 000 individual firms, the average number of employees is around seven per firm. Firms employing seven or less accounted for over a quarter of all work done and almost half of repair and maintenance (see Table 2.2).

Of the work done by contractors, under 60% was done by general contractors, the balance being in the hands of specialist trades. However, as would be expected, these specialised firms had a higher percentage of the repair and maintenance market (see Table 2.3).

Of the specialist trades, the two that stand out are the heating and ventilating contractors, with 6.1% of all work, and electrical contractors, with 8.7%; the balance is spread through a further 17 trades.

As with the earlier analysis of new work and repair and maintenance by size of firm, so the same can be done by trade (see Table 2.4).

Table 2.3 *Breakdown of construction firms, by specialisation, 1991*

	New work, %	Repair & maintenance, %	All work, %
Builders	29.2	33.5	30.8
Builders and civil engineers	26.1	7.6	19.4
Civil engineers	10.0	5.7	8.4
Main contractors	65.3	46.8	58.6
Specialist trades	34.7	53.8	41.4
Total	100.0	100.0	100.0

Source: Annual Housing and Construction Statistics, *Table 3.9.*

Table 2.4 *Analysis of new work and repair and maintenance, by trade*

Size of firm	General builders, %	Building and civils, %	Civil engineers, %	All main trades, %	Specialist trades, %	All trades, %
1–7	33.1	3.4	6.5	19.4	36.5	26.5
8–114	33.8	16.3	26.5	26.9	35.1	30.3
115–299	12.3	9.3	15.5	11.8	8.8	10.6
300–1199	14.7	29.5	39.8	23.2	10.2	17.8
1200+	6.1	41.5	11.7	18.6	9.5	14.8
Total	100.0	100.0	100.0	100.0	100.0	100.0

Source: Annual Housing and Construction Statistics, *Table 3.10.*

*T*he contractor as developer

The growing importance to the industry of 'commercial' construction, or the property boom as it is more colloquially known, has been mentioned earlier; its importance has been enhanced by the extent to which the contractor became his own client, or developer. There is no overwhelming logic to combining the contracting and development functions; however, so many contractors have become developers of one form or another that development must be regarded as an integral part of most large contracting businesses and, hence, a necessary part of this book. At its simplest, the contractor supplies an agreed product where the price, delivery time and specifications have all been determined in advance. His risk is that he will not perform technically as required, or that costs will not be in line with estimates. Again, in its simplest form, the developer commissions work without a final customer, ordering the house, office or factory in anticipation of a tenant or purchaser. His risk is the obvious one that he may not be able to sell the product; if the developer is also his own contractor, he will additionally carry the contracting risk.

WHY SHOULD THE CONTRACTOR BE
HIS OWN DEVELOPER?

Some of the reasons typically given by contractors to rationalise their development activities are justification after the event. However, if development is regarded as an entrepreneurial function, then there is no reason why a contractor should be precluded from exercising that function. Indeed, because of the nature of his work 'on the ground' in a whole variety of geographic locations, frequently for clients who are developers in their own right, the contractor can be singularly well-placed to identify these opportunities. Having identified them, he has the necessary technical skill to execute the project; indeed, he can even be said to be creating his own workload in a way which is open to few other industries.

One can advance various theories as to how and why the contractor became drawn into development. In *Partners in property* (1964) Brian Whitehouse argued that the credit squeeze in 1957 played an important part in bringing the contractor into closer contact with the developer. Among the possible sources of finance:

> *were the very builders and contractors with whom they worked in close co-operation anyway. To these firms there was still available, through the normal channels, finance in relation to the size of contracts on hand . . . moreover, it was well-known that some of the better-known names in the building and contracting world were interested in a small way in property development and ownership on their own accounts . . . what followed was inevitable . . . it was during this period that the builders and contractors became equity-sharing partners in development work to a greater extent than ever before.*
> (Brian Whitehouse, *Partners in property*, p. 52)

Hedley Smyth mentions the incentive of higher margins:

> *Many contractors were looking for new markets during the early 1950's as workloads declined. A number of contractors were willing to finance the construction for developers to prevent under-production. There were reasons for this. First, they anticipated this as an important new market which would yield above-average profit margins. Second, their low overheads, the use of subcontractors and credit from materials' suppliers meant that they could undertake the financing of their own work without threatening their liquidity.*

(Hedley Smyth, *Property companies and the construction industry in Britain*, pp. 137–8.)

What is clear to those who have talked with contractors when development was fashionable is that, working closely with developers, the contractor becomes all too aware of how substantial appears the development margin compared to the low profit margins secured on contracting. The capital investment required for development is then seen as a natural home for the positive cash flow generated out of contracting. Some contractors have even argued for the retention of completed commercial property so that the growing stream of rental income may act as a counterweight to the fluctuations inherent in the construction business.

Development is essentially the creation of new assets and, in a consistently inflationary environment, monetary returns are enhanced. This either makes the development subsidiaries appear exceptionally profitable or, alternatively, covers up mistakes which might have been exposed in a non-inflationary environment. Success may breed success; it also breeds a belief in one's own entrepreneurial ability. Thus, the consistently high development returns made in the late 1950s and through the 1960s left developers singularly ill-prepared to meet the collapse in values in 1973–74. Those who thought the lessons of that period would last at least a generation then saw the inflationary gains of the 1980s reverse even more dramatically at the end of the decade. Thus, there are housebuilders, either independent or within contracting organisations, who have lost significantly more money in the period 1990–92 than they made in the whole of the previous decade.

Although there is a clear economic separation between the development and the construction functions, the extent to which they are in practice conducted separately will depend on the relative strength and expertise of the ultimate client as compared with those servicing him, and the extent to which the latter can anticipate the needs of the former. To take an extreme, it would be unusual to find speculative developments of a brewery or pharmaceutical factory as the developer could not match the product knowledge of the end-user, nor easily anticipate his demand requirements. Thus, speculative commercial development is concentrated on standard factory, office and retail units which have the capacity to suit many occupants, and where the developer can rely on his assessment of overall market conditions and trends, rather than the detailed needs of specialist occupiers. A compromise is often found where the developer will acquire the site and then seek to negotiate the construction of individual units with end users who did not have the expertise to carry out the site assembly.

In housing development, the ability of the end buyer to organise his own site purchase and commission a contractor is, for most of us, non-existent. The individual house purchaser suffers in particular from the indivisibility of one of the most important raw materials – the land. Although individual sites are occasionally available, there is usually a requirement to manage a large site, secure planning permission, and install all the services (drains, roads, power, etc). Although there is no reason why it should not exist over here, we in Britain do not have the American practice of developing large sites into serviced plots for sale to individual purchasers. The development function in housebuilding is, therefore, particularly difficult for the end buyer to handle himself and it does need a development intermediary. Although these exist in the form of specialist housebuilders, the contractors can legitimately claim that their traditional skills of site organisation and construction cover most of what is required. Indeed, it is not always easy to distinguish the contractor integrating backwards into land assembly, from the developer who organises his own construction process (unless you look at the structure of the Board!).

To the extent that contracting experience and opportunism can legitimise diversification into development, then property and housing are not the only routes which can be followed. For instance, contractors' earth-moving skills had taken them into open cast coal mining as contractors to the National Coal Board. These skills have then been exported abroad, first as contractors and then as mine owners; Costain provides the most obvious example. The competition for civil engineering work has also reintroduced the old concept of the nineteenth century contracting railway promoter. The most notable example is the British and French contractors' promotion of the Channel Tunnel, followed by the proposals for a high speed rail link to London. Examples can also be found in road construction; John Laing was an investor in Spanish motorways in the 1970s, and contractor equity has helped finance road bridges such as the Dartford Crossing and the Severn Bridge. It is unfortunate that such entrepreneurial opportunities are presenting themselves at a time when the contractors' capital base has been savaged by earlier ventures into property development.

MORE ABOUT THE HOUSEBUILDER

The structure of the individual housebuilder is not complex. They buy a little land, get some sub-contractors to build houses on it, sell those houses to the likes of you and me, and then take 20, 30 or perhaps even 40% gross margins. Most people, if asked to set up a chemical complex or organise a cement works or even, perhaps, the proverbial brewery, would not know where to start. However, if presented with a million pounds and told to 'start a housebuilding operation', then they might feel that they could at least make a rough stab at it. On such principles have fortunes been made – and lost. It looks simple and when house prices are rising strongly, then even the most incompetent can make money, which is not necessarily true of most industries.

Where is the money made?

There is no doubt that an important influence on profitability is the rise or fall in house prices and their consequent impact on land values. Dealing with this in terms of rising prices, which has been the norm for most of the post-war period, the inflationary profit is made during the time that is needed to acquire a site, to organise the detailed planning requirements, to build the home and to sell it. For most businesses the stock ratio is the number of times that the stock can be turned over in any one year. With housebuilders, it is more a case of looking at the number of years it takes to turn the stock once, or at least the land element of the stock. Thus, it is self-evident that if the housebuilder is buying land on the basis of house selling prices of £50 000, and before he has brought the house to market the price rises to £60 000, then that extra £10 000 profit comes straight through to his bottom line solely as a result of inflation. In a bull market, some housebuilders were buying land on the basis of a projected continuing inflation (even though they all denied it at the time) and that of course only exacerbated the problems of the downturn when it finally came.

There is a minimum period for which the housebuilder has to hold his land in order to physically carry out all the necessary processes, both of planning and construction, and few housebuilders will tell you that they are relaxed about carrying a land bank of less than two years in duration. They can, of course, extend that voluntarily by holding a long-term land bank and there are a number of examples in the lists of housebuilders who have a five to ten year land bank. There has been a fundamental difference of view between those companies that in part regard themselves as land bankers, e.g. Wilson Connolly and Prowting, and those which have regarded themselves as current traders, e.g. Berkeley Homes.

The land profit does not depend on inflation

It is sometimes said by outsiders that the land profit made by a house-builder is all inflationary and, in a period of zero inflation, the house-builder would earn no more than a normal building margin, say, 3, 4 or 5%. Thus, any excess over this is purely a reflection of inflation on land values. This does, of course, ignore the fact that there is a financing cost to be recouped through profit margins, but this somewhat puritanical approach to the housebuilder fails to recognise that they are also making a non-inflationary profit from land development.

The acquisition and development of housing land is a separate and legitimate economic activity in its own right and one which should earn its own reward. The highest returns from land development are made where the housebuilder acquires land which has no planning permission, at a price which is nearer to its alternative use value, and with the housebuilder successfully working the land through to full residential land use. This may in part be a risk return to a speculator's acquisition of assets or, alternatively, a return to the labour and skill involved in convincing the planners of the social desirability of giving consent to one piece of land rather than another. There will also be returns to be made from acquiring land with permission to build at one density, e.g. five per acre, and negotiating a higher density, say, ten per acre. However one regards the ethics of returns made from land assembly and the planning process, they are patently a return for skill and expertise, and quite distinct from any return produced by inflation.

Even when purchasing land with full detailed planning approval, the housebuilder will still generate a legitimate development profit. In this role he is both a retailer of land and a provider of services. The end purchaser wants one plot of land, typically a tenth of an acre, and land does not come on the market in those sizes. It comes on the market in blocks of two acres, five acres, ten acres or whatever, and it requires a wholesaler, or developer, to purchase it in bulk and to subdivide it into the form in which it can be purchased by the individual occupier, albeit with a house on top. In addition, he is fully servicing the raw land which he buys by landscaping the site, putting in the access roads, drainage and all the other services – water, gas, electricity, etc. Thus, within the overall land profit can be found the inflation element (more recently a loss), the return for securing the best planning permission, the land servicing profit, and the wholesaling profit.

The organisation of a housebuilder

The organisation of a typical housebuilder will encompass land buying, construction, and selling, with perhaps the administrative/finance role tacked on to the side. The land buying is probably the most important aspect of the housebuilder's business and the specialist housebuilder will often have this as a Main Board function. Indeed, sanction for land purchases may sometimes rest with the Managing Director himself. The old retailing cliche of 'location, location, location' can just as easily be applied to the housebuilding industry. Land may be bought at auction but most housebuilders prefer private negotiation. The sellers would typically be farmers and public bodies, although economic necessity has added other housebuilders and receivers in more recent years. Land is either bought with or without outline planning consent, that is, where the principle of residential land use is accepted. Detailed planning consent, i.e. covering density and design, still needs to be negotiated, and even this can be quite time-consuming.

Land without outline consent may be in an area which is zoned for residential use, land at the bottom of your garden for instance, but this example alone will indicate that permission may or may not be obtainable according to the local circumstances. This type of land acquisition or infilling is usually for very small sites and most of the medium to large developers will not be active in that part of the market. They will be more interested in land which currently lies outside the residential zonings. County Structure Plans must be drawn up every five years and these stipulate what land use the Council regards as acceptable over the next five year period. The housebuilders' skill is in owning land which is currently outside the structure plan but which will be included in the next review. The housebuilder will also use his professional skill to encourage the planning authorities to include his specific land in the structure plan.

Housebuilders are not over-keen to pay for white land (land not zoned for residential use), certainly not much more than agricultural value, in which case the potential seller is not usually interested in dealing at that price. The housebuilder will sometimes buy at a relatively low value with the condition that there will be a substantial extra payment if planning permission is obtained. More frequently he will take an option on the land and use his skill and capital to see that land through to successful planning. If he achieves planning, then he will pay the option price for that land. That may be a fixed rate per acre or, more commonly, it will be expressed as a percentage of the market price. Once, the discounts were in the 25–50% of market value range, but a combination of growing sophistication on the part of the sellers and their advisers, and growing competition from the buyers, meant these discounts gradually narrowed to 5–10% and occasionally even reached market price, on the grounds that such a deal avoided the housebuilder having to pay an excessive price at auction. During 1990–92, with a different relationship between demand and supply, more sizeable discounts began to emerge again.

Moving on from the land assembly, the building process is more uniform. The developer does not normally let the whole of the building contract to a builder (which would involve building well in advance of short-term sales rates); neither does he carry out the totality of it with his own labour. Normally he will employ supervisory staff, particularly the site agent, and perhaps also specialist groundwork staff, and will sub-contract out the specialist trades, such as bricklaying, plastering, painting, roofing, etc. The building will be carried out in phases to limit the amount of capital employed, which is easier to do on low-rise housing than it is on a large block of flats which have to be built in their entirety. When contractors expanded their housing operations, they tended to regard it as an offshoot of construction, which is never a successful approach. Land was often treated as a commodity to be bought like the bricks, and with the same subtlety, and the construction was carried out in-house, lacking the sharpness of external pricing. The contractors also tended to be relatively capital intensive because they would typically put in all the infrastructure at the beginning of the site as it offered the best construction solution, rather than to put in the roads parallel with the rate of sales progress.

The final organisational heading mentioned was the selling side of the business which merits only brief comment. Aggressive selling and marketing tends to be a more cyclical affair than in most other industries. In the good times, the office is opened, the orders are taken and frequently customers will buy off plan rather than waiting to see the finished product. Supply is limited by decisions taken months earlier, so excess marketing serves no short-term purpose. In contrast, in the recession, the selling effort becomes more intensive of labour and a whole range of selling aids are dusted off: mortgage subsidies, free legal fees, part-exchange, stamp duty, etc. We will meet these 'goodies' again when examining accounting policies in Chapter 9.

Construction statistics: sources, relevance and problems

DoE OUTPUT AND ORDERS SERIES

One can define the construction industry in a variety of ways, accompanied by interesting semantic arguments on where construction stops and where engineering, or property development, or whatever, begins. Alternatively, we can turn to the sponsoring ministry, the Department of the Environment, who might well have said that 'when I define the construction industry, it means just what I choose it to mean, neither more nor less.' The Department produces the regular statistics of construction output and orders, and housing starts and completions, on which all independent forecasts are based; for practical purposes it does define the construction industry for the outsider. What is important for the analyst is to know what is and what is not included, not only within the total definition of the industry but also within the sub-sections – all is not what it sometimes appears.

An important caveat must be introduced before the main body of statistics is described. In November 1992 the DoE altered the definition of certain categories within the New Orders statistics, recognising the

Table 4.1 *1991 Construction output and orders, current prices*

	£m		%		£m		%
			Output				*Orders*
New Housing – Public	810		3.2		875		4.5
Private	4 846		19.4		4 552		23.4
		5 656	22.7			5 427	27.9
Other new work –							
Public	5 772		23.1		4 767		24.5
Private industrial	5 314		21.3		3 452		17.7
Private commercial	8 224		32.9		5 811		29.9
		19 310	77.3			14 030	72.1
Total new work		24 966	100.0	57.1		19 457	100.0
Repair and maintenance							
Housing – Public	3 964		21.2				
Private	5 804		31.0				
Other work – Public	4 807		25.6				
Private	4 168		22.2				
Total repair and maintenance		18 743	100.0	42.9			
Total all work		43 709		100.0			

Source: Housing and Construction Statistics Great Britain, *Tables 2.3, 2.6, Sept. 1992.*

changes in ownership consequent on privatisation, and creating a new infrastructure category. Although ten years' comparative back data was published at the same time, the more detailed breakdown of the data is not yet available. The output series continues to be prepared and published on the old basis and, at the time of writing, it is not expected to be issued on a revised basis until late 1993. This section takes the old format, which runs continuously from 1955, as the starting point, using for illustration the 1991 annual figures (see Table 4.1), these being the last available on a comparable basis; then follows a more detailed description of the changes now being introduced.

Availability

The Output and Order series are published in the quarterly and annual *Housing and Construction Statistics*; there are also press releases, quarterly for output and monthly for orders. They are published both in current price form (i.e. in the values of the day), and in constant prices, presently deflated to 1985 equivalent prices.

Definition of output

The amount chargeable to clients in each quarter is taken as the output. Contractors would include work done on a speculative basis for themselves, thereby covering large parts of the private housebuilding industry.

Unrecorded output

The output series includes estimates for unrecorded work done by firms and individuals not on the statistical register. As these estimates are based on the number of self-employed construction workers published by the Department of Employment, the full scope of 'black economy' construction output is not recorded. Probably more important is that construction work carried out by firms or individuals who are not them- selves construction companies is unrecorded. The construction statistics should therefore be regarded as relating to the output of contractors rather than of all construction. The only exception to this is that the construction output (commonly known as direct labour) of government departments and local authorities is included. Excluded is the work done by most nationalised (or now privatised) organisations, including gas, electricity and water, construction work carried out by private firms on their own account, and the do-it-yourself activities of the public. This own-account work will be concentrated on the repair and maintenance categories, and those serving the construction market, e.g. building materials producers, may find that the DoE construction output stat- istics do not wholly represent the available market.

Constant price data

Because of the varying impact of inflation, it has become standard practice to use the constant price series of output and orders, almost to the point where the existence of the current price series is forgotten. In producing a constant price series (presently based on 1985 prices) it follows that a price index must be used to deflate the actual statistics collected; adjusting for estimated changes in prices has the capacity to introduce error into the constant price series. The output price indices now used were introduced in 1978 with adjustments back until 1970. The price deflator applied previously was considered suspect: 'The reliable measurement of price changes for construction work poses severe problems and it is doubtful whether these are surmounted satisfactorily by the index used officially for deflating current output values.' (*Construction and the related professions*, M.C. Fleming, 1980, p. 131; written as the new indices were being introduced.)

An explanation of the methodology of the price indices is contained in the Appendix to the *Annual Housing and Construction Statistics* with a fuller description in *Economic Trends*, No. 297, July 1978: A.D. Butler, 'New price indices for construction output statistics'. In brief, the output price indices are derived from tender price indices prepared by the Property Services Agency, the Department of Transport and the DoE, with separate indices applied to each of the five main categories of material and labour. The output index for a given quarter is based on the proportions of that quarter's output which had been secured in earlier quarters, and the tender price indices ruling in those earlier quarters.

The output price indices are, therefore, only as good as the tender price indices; the appropriateness of the specific tender price index to the class of output being measured, and the allocation of work over time. There has been concern over the accuracy of the constant price output series in reflecting the onset of recession in 1990. Throughout 1990, the constant price series recorded year on year percentage increases, yet during that time the leading building materials statistics were showing double digit percentage declines. The DoE explanation is that the high volume of finishing trades involved in the commercial building boom (e.g. air conditioning, painting) were still feeding through to output, whereas the building materials for which statistics are available tend to be used earlier in the construction process. Figure 4.1 shows the percentage change in cement deliveries and construction output, based on a rolling four quarterly period for smoothing. The movements in cement deliveries can be amplified by stocking and imports but the two lines remained very close (as one would expect) between 1975 and 1988; however, since then, construction output has fallen less sharply than cement deliveries.

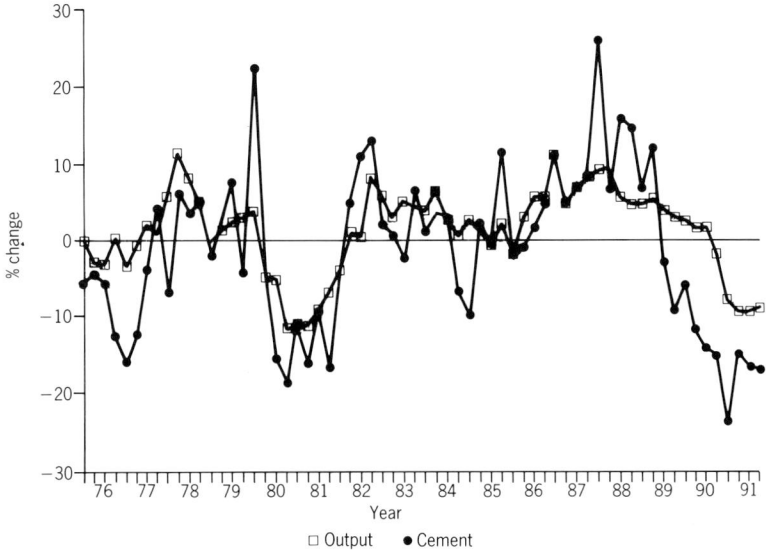

4.1 *Construction output* vs. *cement deliveries.*

Definition of sub-sectors

Housing, public

All local authority and housing association housing, plus site work and services.

Housing, private

All privately owned residential buildings, plus associated site works.

Public non-housing

This is a category which has been substantially changed in recent years by privatisation. There is a distinction between the effect on orders and on output. The order is placed in the public or private sector according to the status of the order-giver at the time, i.e. once privatised, all orders go into the private sector. However, the output is attributed to the public or private sectors according to the status of the original order. Thus,

Table 4.2 *1991 Public non-housing current prices*

	Output		Orders	
	£m	%	£m	%
Gas, electric, coal	250	4.6	59	1.2
Railway, air transport	331	6.1	252	5.3
Education	537	9.9	734	15.4
Health	578	10.6	578	12.1
Factories, warehouses	130	2.4}	665	13.9
Offices, shops, garages	523	9.6}		
Roads	1487	27.3	1386	29.1
Harbours, waterways	207	3.8	133	2.8
Water	37	0.7	16	0.3
Sewerage	195	3.6	202	4.2
Other	1166	21.4	743	15.6
Total public non-housing	5442	100.0	4768	100.0

Source: Housing and Construction Statistics Great Britain, *Tables 2.4, 2.7, Sept. 1992.*

after privatisation, output only gradually switches from the public to the private sector.

Table 4.2 shows the relative importance of the component parts of public non-housing construction (the output split has only recently become available).

Most of the categories speak for themselves but the existence of factories, offices and shops within the public sector deserves comment: they include all local and central government offices, vehicle garages and workshops, bus depots and shopping developments sponsored by local authorities. The miscellaneous category is largely security related, including all defence spending, prisons, police and fire stations.

Private industrial

Historically, this largely comprised factory building for manufacturing industry and warehousing. In the 1970s there was the addition of North Sea oil investment, followed in the 1980s by the privatised utilities and, for good measure, the Channel Tunnel in July 1987. The comments made earlier about the different treatment of output and orders for the privatised industries applies to the industrial sector. Table 4.3 below shows the extent to which the privatised industries have distorted the

Table 4.3 *1991 Industrial current prices*

	Output		Orders	
	£m	%	£m	%
Factories	2290	43.1	1832	53.1
Warehouses	642	12.1	438	12.7
Water }			494	14.3
Sewerage }	2383	44.8	227	6.6
Electricity }			255	7.4
Other }			206	6.0
Total	5314	100.0	3452	100.0

Source: Housing and Construction Statistics Great Britain,
Tables 2.4, 2.7, Sept. 1992.

sector; the new series, discussed later, will return the industrial sector to its earlier roots.

The effective dates for the larger transfers between sectors were:

Nov 1984	British Telecom
Dec 1986	British Gas
Feb 1987	British Airways
Jul 1987	BAA
Dec 1988	British Steel
Dec 1989	Water Authorities (England & Wales)
Dec 1990	Electricity Distribution (England & Wales)
Mar 1991	Electricity Generators (England & Wales)

Private commercial

Running through the private commercial sector is speculative property development, which will underlie the largest categories of offices and shops and perhaps even entertainment. The balance between the sub-sectors is shown in Table 4.4.

As with industrial, the individual categories are largely self-explanatory; the 'others' category includes private airfields, nursing homes, clinics, churches, et al.

Table 4.4 *1991 Commercial current prices*

	Output		Orders	
	£m	%	£m	%
Offices	4318	52.5	2216	38.1
Shops	1446	17.6	1222	21.0
Entertainment	1096	13.3	1111	19.1
Garages	342	4.2	261	4.5
Education	173	2.1	169	2.9
Agriculture	116	1.4}	833	14.3
Others	733	8.9}		
Total	8224	100.0	5812	100.0

Sources: Housing and Construction Statistics Great Britain,
Tables 2.4, 2.7, Sept. 1992.

Repair and Maintenance

No official description of the content of the repair and maintenance category, as distinct from the new work, exists. The difficulty lies in the definition and treatment of work which 'improves' rather than merely maintains an asset. For housing, all improvement work, however substantial, is categorised as repair and maintenance so the distinction is not of importance. However, for non-housing, improvement work is treated as new output rather than repair and maintenance. It is left to the contractors to report work in the category they believe appropriate, which leads to the inevitable subjective judgements: how thick a layer of tarmac is needed to turn repair and maintenance into improvement? The relative importance of the different categories of 'repair and maintenance' are shown in Table 4.5. (There are no equivalent order statistics.)

Little information is available on the mix of work within the categories in Table 4.5. Economically, the work ranges from small revenue items, typically met out of current income, up to large projects which require significant capital commitment, often financed by borrowing. From time to time the concept emerges that repair and maintenance is unaffected by recessions; this is a fallacy which stems from the 'Repair and maintenance' categorisation (see Chapter 10, which covers forecasting). 'Public housing' ranges from minor maintenance for tenants to multi-million pound contracts for what amounts to complete reconstruction of inner-city areas. 'Private housing' encompasses anything from decorating

Table 4.5 *1991 Repair and maintenance output current prices*

		£m	%
Housing	– Public	3 964	21.1
	Private	5 804	31.0
Other Work	– Public	4 807	25.6
	Private	4 168	22.2
Total		18 473	100.0

the bedroom (if done by a registered contractor) to new drives, re-roofing, or even extensions. The most publicised area of 'public non-housing' is road maintenance but, to the extent that it is regarded as improvement, it will be contained in non-housing new work. For both the public and private sectors, the repair and maintenance will largely comprise the good housekeeping of the multitude of offices, factories, town halls, hospitals and so on.

The New Series

A new 'Infrastructure' sub-sector has been created. This comprises water, sewerage, electricity, gas, railways and harbours, taken from public non-housing and private industrial, and roads and air transport from public non-housing and private commercial. The relationship between the old and new series for the non-housing new work sectors is shown in Fig. 4.2. The residual public non-housing sector is around half its pre-privatisation levels. Commercial building is more or less the same in both series until 1991, but Fig. 4.2 shows the extent to which industrial work had been distorted since the mid-1980s by projects which were once the remit of the public sector. The housing and repair and maintenance sectors are unchanged.

HOUSING STATISTICS

There is an extensive body of statistics covering the housing market. Those discussed here relate to activity levels: transactions and new

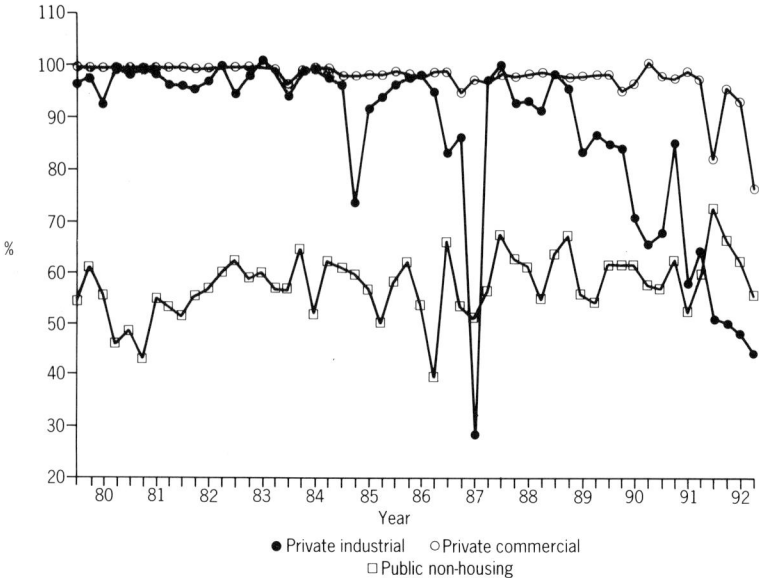

4.2 *The relationship between the old and new series for non-housing new work sectors.*

building. There are other aspects of the housing market where the statistical data is extensive, complex, and frequently misinterpreted. For this reason there are separate chapters on household formation and the long-term demand for housing, and on house prices and affordability.

Gross housing sales

Although precise figures are available for the number of new houses physically completed, there is no definitive series for houses sold, either new or old (a series for new houses completed but unsold did exist but was stopped at the end of the 1970s). However, there are two series which can be used to obtain a rough indication of the magnitudes involved in gross housing sales: mortgage statistics, and the Inland Revenue property transaction series.

Table 4.6 *Number of mortgages: United Kingdom (000)*

	1988	1991
Building Societies	1232	697
Banks	357	315
Insurance companies	41	13
Local authorities	11	4
Total	1641	1029

The figures are mortgage completions, except for banks which are mortgage approvals.

Mortgages

Most, though not all, houses are bought with the assistance of a mortgage and the main lenders publish figures for the number of transactions. Those for 1988, the peak year, and 1991, the latest year for which there is complete information, are shown in Table 4.6.

Inland Revenue series

This is a new series first published in *Economic Trends* in June 1991, and backdated to April 1977. The series is based on the property transactions notified to Inland Revenue Stamp Offices or District Land Registries in England and Wales. There was a detailed explanatory article accompanying the series but those wishing to consult it should be aware that parts of the text and tables suffered from gremlins and were revised and reprinted in the July issue.

A gross figure for the total number of property transactions is calculated on a monthly basis and this inevitably becomes the 'headline' figure, taken as shorthand for what is happening in the housing market. However, the gross number of transactions also includes land and non-residential transactions and a more detailed analysis of the data is provided on an annual basis (see Table 4.7). This shows that the residential content tends to be around 92–93%. A small proportion of those residential transactions will be by corporate bodies (3% to 4%). Thus, of the headline figure announced each month, individual residential transactions account for around 90%.

Table 4.7 *Inland Revenue property transactions: England and Wales*

	1988	*% of Total*	*1991*	*% of Total*
Total	2148	100.0	1315	100.0
of which residential	1990	92.6	1225	93.2
of which individual	1927	89.7	1190	90.5

The Inland Revenue series is published in Economic Trends *and reprinted in* Housing Finance.

Forms are required to be returned within 30 days of the transaction and there is therefore a lag of around one month in the series. It should be noted that the Inland Revenue figures are for England and Wales whereas the mortgage figures are for the UK.

Housing starts and completions – DoE

Whereas the Inland Revenue series cover transactions in both the new and second-hand markets, the DoE statistics relate to the construction of new dwellings. Prior to 1980, monthly figures were based on the returns received, which could include activity relating to earlier periods (believed to be small). Since January 1980, the housing statistics relate to the activity within the period and are therefore subject to revision for late returns. There is extensive classification by region, type of dwelling and tenure, contained within *Housing and Construction Statistics*, but the monthly press release, the most topical source, analyses according to private enterprise, housing associations, local authorities, new towns, and government departments. The latter two are normally grouped and referred to as the public sector but the categorisation of housing associations as between public and private is becoming more difficult. (For connoisseurs of Government definitions, a house is 'a dwelling which is not a flat.')

A 'start' is the date when work begins on the foundations or slabbing (but not site preparation); this does not necessarily represent the moment when the individual dwelling unit actually begins. The 'completion' is the date on which notice of completion or occupation is given to the local authority under building control regulations. Where houses remain unsold and the possibility of paying rates or community charge on empty property arises, there may be some temptation to delay final completion.

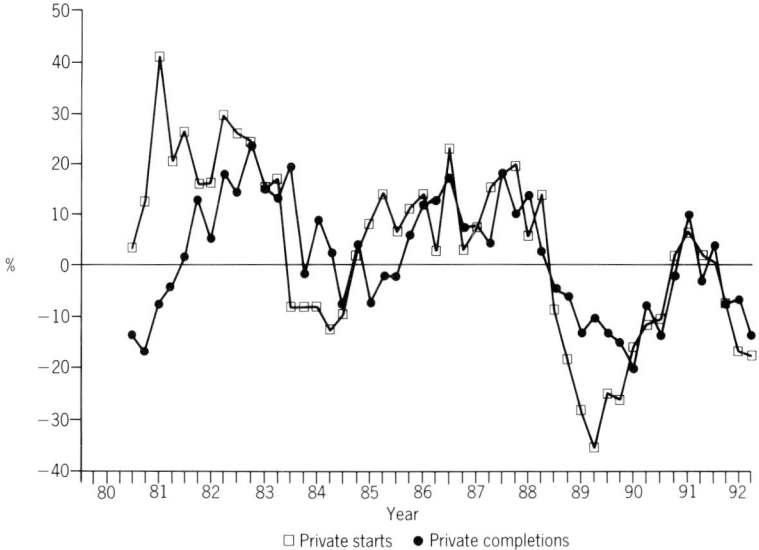

4.3 *Private housing starts* vs. *completions.*

Indeed, it is not unknown for completions to respond coincidentally with starts, despite construction logic dictating a delay. Figure 4.3 shows private housing starts against completions.

Private housebuilding statistics – NHBC

The National House-Building Council operates the registration scheme to which most private housebuilders belong; this is the one which offers two and ten year warranties to purchasers. Until 1989 the NHBC operated the only registration scheme but at that date Foundation 15 (a subsidiary of Municipal Mutual Insurance) began an alternative scheme; this may have taken 3–4% of the market until it ceased accepting new business in late 1992 (at the time of writing, however, there are plans to restart in August 1993). The NHBC series also misses those builders who choose not to register at all, and the self-build organisations; the NHBC estimates it has a coverage of around 90% of the private sector. The NHBC statistics additionally include housing associations who choose to register with the Council.

Because the NHBC 'starts' are in fact registrations, which must be made at least 21 days before building begins, they can give an earlier indication of building activity than the DoE figures; however, although the figures are compiled monthly, they are published quarterly.

Housebuilders' Federation

During 1992 the Housebuilders' Federation began surveying a sample of housebuilders on such sensitive statistics as site visitors, net reservations and expected sales over the coming year, the results being published in *Housing Market Report*. As yet it is too early to assess its value but it already appears to be emerging as one of the leading indicators of the state of the private housing market.

Which housing indicator?

There are a number of indicators which the analyst could legitimately use as a guide to activity in the housing market and it is interesting

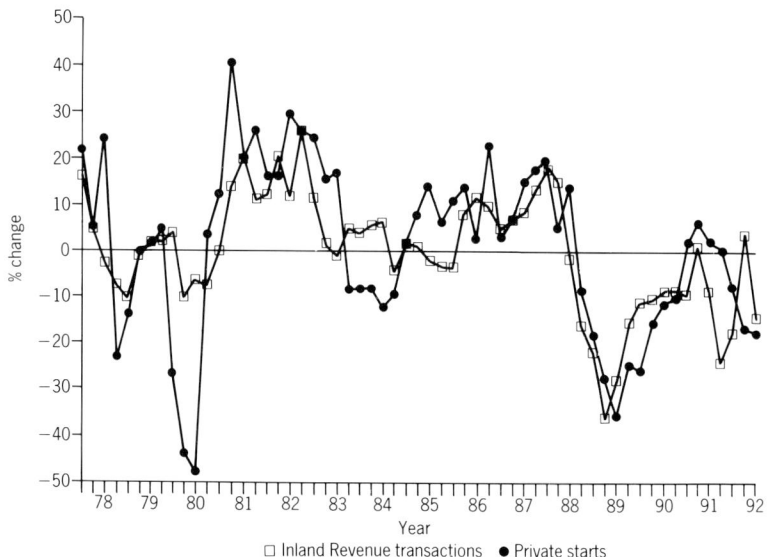

4.4 *Inland Revenue transactions* vs. *private housing starts.*

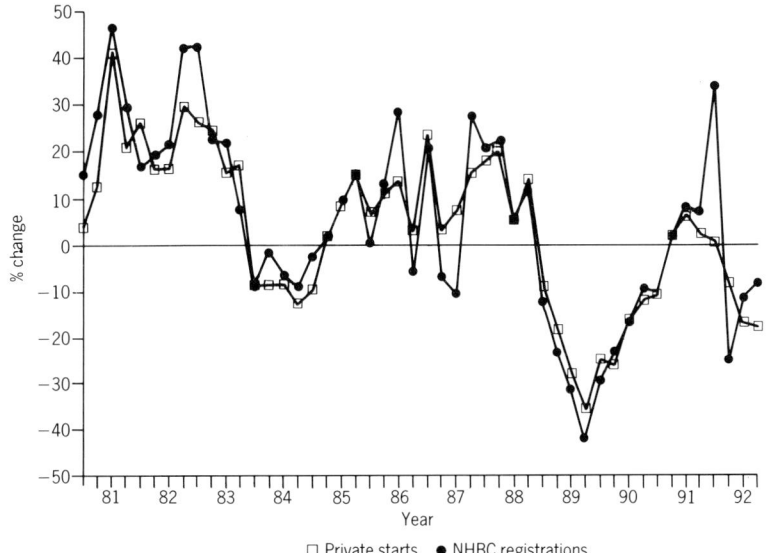

4.5 *DoE starts* vs. *NHBC registrations.*

to compare the timing and extent of the cyclical patterns. Figure 4.4 compares the changes in the Inland Revenue series with private housing starts and there is little evidence of leads or lags. Indeed, with the exception of the near collapse in starts in 1980, the size of the peaks and troughs is surprisingly similar; one might have expected new building to amplify the changes in total transactions. Figure 4.5 plots the changes in DoE starts and NHBC registrations which match each other very closely; the upsurge in registrations which took place in isolation in the first quarter of 1992 was influenced by the fee increase scheduled for April.

NON-HOUSING NEW WORK

Industrial

Chartered surveyors King, Sturge & Co. prepare *Industrial Floorspace Survey*, a publication which shows their estimate of the total industrial floorspace available (not necessarily empty) for letting. There is a regional analysis and an estimate of newly constructed space. The data, which is published three times a year, carries back to April 1975. Another important indicator of industrial investment is the CBI's *Industrial Trends Survey*. That specifically asks respondents whether they expect to authorise more or less capital expenditure over the following twelve months, on buildings and plant and machinery separately. The answer is normally expressed as a balance of those expecting to authorise more over those expecting to authorise less, without any magnitudes. Manufacturers are also asked whether or not they are working below capacity and a similar balance figure can be calculated. Both of these series are held on Datastream, back to 1979 for investment in buildings and 1972 for capacity utilisation. These series are discussed further in Chapter 10 on forecasting.

Commercial

Several of the larger commercial estate agents maintain research departments which produce regular bulletins; these include a profusion of statistics relating to the state of the market from which trends can be taken. The statistics are based on personal observation and experience rather than official returns, but are none the worse for that. The data is generally confined to the London market, and usually split between the West End, City and Docklands. Table 4.8 below shows a typical selection on London offices from Jones Lang Wootton, aggregated for the whole of Central London.

Other detailed research on the commercial property market is now provided by Applied Property Research (of 97 St. John's Street, London EC1) but this is only available on special subscription.

RENOVATIONS

Housing and Construction Statistics contains several tables on renovation work but they can be amongst the more challenging to use. For the public sector, the number of dwellings and the value of the work

Table 4.8 *Central London office space, units over 5000 sq. ft. (sq. ft., 000)*

	Take-up of space	Development starts	Completions*	Availability†
1983 1st half				
2nd half	3 900			
Year		3 680	3 910	8 840
1984 1st half	4 200			
2nd half	3 950			
Year	8 150	4 160	3 850	8 310
1985 1st half	5 920			
2nd half	5 910			
Year	11 830	4 520	4 140	7 550
1986 1st half	6 570			
2nd half	5 725			
Year	12 295	4 880	4 150	6 140
1987 1st half	6 450			
2nd half	6 300			
Year	12 750	13 790	4 650	7 570
1988 1st half	5 010			
2nd half	4 070			
Year	9 080	10 700	7 240	13 500
1989 1st half	4 440			
2nd half	4 580			
Year	9 020	8 510	7 790	216 400
1990 1st half	4 100	4 100		
2nd half	3 790	2 865		
Year	7 890	6 965	9 290	25 780
1991 1st half	2 400	1 330		
2nd half	2 725	830		
Year	5 125	2 160	12 570	2 578
1992 1st half	2 860	330		
2nd half	3 160	180		
Year	6 020	510	5 730	24 385
1993 est		1 340	3 010	
1994 est		2 570	1 578	

Source: Jones Lang Wootton Research.
* *Includes owner occupied and pre-lets.*
† *Vacant or available for occupation over the next twelve months.*

done is recorded; for the private sector, the data relates to the number and value of the grants made, but there is no measure of the value of the work ultimately carried out under the grant. Thus, only the total number of dwellings covered can be aggregated, though with no distinction between timing or size of project.

49

Table 4.9 *Number of renovation grants or dwellings: England*

	1992	
	000	
Private sector grants		
1989 Act Mandatory grants	47.0	
Discretionary grants	5.6	
Minor Works Assistance	29.3	
1985 Act All grants	3.9	
Total		85.8
Public sector works completed		
Housing Associations	7.2	
Local authorities and New Towns	175.3	
Total		182.5
		268.3

Source: Housing and Construction Statistics Quarterly,
Table 2.22. Tables 2.15 to 2.21 of Housing and Construction
Statistics Quarterly *give more detail for the categories above but, for
grants under the 1989 Act, not always in a way which is consistent with
the summary table.*

The tables of work done for local authorities and housing associations have been consistent over the years, but those for the private sector have varied significantly according to the legislation under which the grants have been made available. Currently, the *Housing and Construction Statistics Quarterly* has renovation tables from 2.15 to 2.22, the last being close to a summary table. The lists renovations by region; as figures for Scotland are not available for all categories, the composition of the English renovations are shown in Table 4.9.

The private grant scheme was introduced in 1949 but there have been extensive changes in the types of grant, their amounts, and the accompanying regulations as the Government seeks to match contemporary needs with contemporary financial resources. Moving back chronologically, significant changes have been introduced in 1989, 1985, 1980 and 1974; Table 4.10 gives a summary of the number of dwellings subject to public sector or grant aided renovations over the last two decades.

The statistics for the 1970s show renovation work behaving much as new housing, peaking in 1973 before a collapse and then a slow recovery

Table 4.10 *Renovations in England (000)*

	1971	1972	1973	1974	1975	1976	1977	1978	1979	1980
Local authorities	59.1	97.5	110.1	73.5	36.2	39.0	37.6	60.9	76.0	77.3
Housing Assocns.	5.0	4.2	3.2	4.0	4.6	13.4	18.8	13.1	17.2	14.7
Public sector	64.1	101.7	113.3	77.5	40.8	52.4	56.4	74.0	93.2	92.0
Private grants	90.9	124.2	166.0	192.3	85.4	68.7	57.0	57.6	65.4	74.5
Total	155.0	225.9	279.3	269.8	126.2	121.1	113.4	131.6	158.6	166.5

	1981	1982	1983	1984	1985	1986	1987	1988	1989	1990
Local authorities	52.9	57.7	85.5	86.6	96.5	133.7	148.4	169.0	194.9	226.5
Housing Assocns.	11.2	17.4	14.5	18.5	11.4	12.7	10.9	11.2	13.0	10.7
Public sector	64.1	75.1	100.0	105.1	107.9	146.4	159.3	180.2	207.9	237.2
Private grants	68.9	104.0	219.8	229.1	136.4	113.3	108.9	105.3	98.2	96.7
Total	133.0	179.1	319.8	334.2	244.3	259.7	268.2	285.5	306.1	333.9

	1991	1992
Local authorities	177.3	175.3
Housing Assocns.	6.6	7.2
Public sector	183.9	182.5
Private grants	84.8	85.8
Total	268.7	268.3

Source: Housing and Construction Statistics Quarterly, *Tables 2.15 and 2.22.*

at the end of the decade. The pattern in the 1980s was quite different with public policy clearly emphasising improvement in the housing stock against new build. The private sector was particularly affected by changes to the terms of repair grants in 1982 whereby for a temporary period (later extended) up to 90% of the value of the work could be covered by grant; this gave rise to what was popularly called the 're-roofing boom'.

*H*ouse prices and affordability

HOUSE PRICES

Beware the last decimal point

House prices are amongst the most frequently quoted statistics in the construction industry, ranging from academic theory, through commercial planning and not forgetting bar room discussion; all the more reason that their users should understand the limitations of the different series. This section is not a theory of house price determination but a brief guide to the most widely used house price statistics, with reasons why one series might be preferred to another. The essential point to remember when presented with an authoritative statement that house prices moved last month by an amount calculated down to one tenth of a percentage point is that, like so many economic statistics, the figures should not be treated with that degree of accuracy. There are considerable theoretical and practical problems in calculating any index of house prices and it is not unknown for two of the leading indices to produce substantially different results at any given point in time.

Houses are not a homogeneous product. Even where two houses are of identical construction, they may be located in different areas, or on the wrong side of the tracks, or with a different size plot, or more sunshine, or whatever. The condition of two adjacent second-hand semi-detached houses could vary enormously, or the contents may be different (and priced with an eye to stamp duty savings rather than statistical convenience). Over time, other divergences creep in; for instance, individual leading organisations may target one particular segment of the market, either deliberately or by default. The banks tend to mortgage higher priced properties (although the gap narrowed considerably in 1991) and so their entry and departure from the mortgage market affects the pool of house buyers available for the building societies. Thus, the building society share of the mortgage market fell from 96% in 1977 to 56% in 1982 rising back to 86% in 1984, distorting their average prices relative to the totality of the market. There have also been changes in the proportions of small (cheap) and large (expensive) houses which distort simple average prices. This was accentuated in the days of mortgage rationing by the building societies' practice of giving priority to first time buyers when funds were in short supply.

The house price series that existed up to the 1980s were predominantly based on simple average prices with little, if any, allowance for the change in the mix of houses passing through their books. However, the sharp change in the building societies' portfolio of lending in the early 1980s stimulated the introduction of more sophisticated methods of adjusting for the change in the mix of properties being mortgaged; some of these series have been extended backwards, though not necessarily published officially.

One of the most comprehensive reviews of the principles and methodology of house price indices, *Spon's house price data book* by Fleming and Nellis, (published in 1987) devotes 400 pages to the subject; it includes a description of 18 different indices and summary data for each of them since the commencement of each individual series. This book concentrates on the four series that fall into the category either of being in the official statistics books, or most frequently in the public domain.

a) Department of the Environment/Building Societies' Association – Average Prices

This series is published in the Council of Mortgage Lenders' *Housing Finance* (what was the *BSA Bulletin*); and by the DoE in its quarterly (Part 1, Table 1.11) and annual *Housing and Construction Statistics*. The series is available on a continuous basis from 1956 for new houses and 1975 for all houses and it is based on advances approved. There have been changes in the method of obtaining the returns from the societies, one of the most important being that, since 1981, houses sold at below the market price have been excluded. The main drawback of the series is that it is not adjusted to reflect changes in the mix of properties mortgaged.

b) The Department of the Environment Five Per Cent Sample – Average Prices and Index

This series is based on a detailed sampling of the building societies at the mortgage completion stage; the sample was extended to all lenders during 1992. This series commenced in late 1965 for new and existing houses; regional data and first time buyers were added from 1968. In the early years the index was based on a simple average for all houses, including houses sold at below the market price, and without adjustment for mix. A regional breakdown of average house prices showing the average advance and average income of the borrower is published on a quarterly basis in *Housing and Construction Statistics* (Part 1, Table 1.13). Thus, although comprehensive and of long standing, the series was overtaken by other, more sophisticated, indicators.

However, in 1982, the Building Societies' Association published 'a mix adjusted series', again calculated by the DoE and based on the five per cent sample. In fact, this index had existed for some time and had been used to calculate the mortgage financing element of the retail price index. The new mix adjusted all dwellings series was calculated back to the second quarter of 1968 (and to the first quarter of 1969 for all the regions), published as an index in the BSA's monthly press release; and in *Housing Finance* (e.g. Table 18 in August 1991). The mix adjusted series is also the source of the price index that appears in the opening summary tables of construction costs and prices in Part 2 of *Quarterly Housing and Construction Statistics* (Table 2.2).

Users must have regard to the fact that the data is collected at the mortgage completion stage and that the series is not published as promptly as some; the five per cent sample mix adjusted series is not, therefore, the most sensitive indicator of price changes in the housing market.

c) Nationwide Building Society

The Nationwide Building Society (formerly the Co-operative Permanent) has the longest continuous run of data on house prices, starting in a limited way in 1946 and predating the basic DoE/BSA series by ten years and the next individual building society index by almost thirty years. For continuity alone it demands attention but the Nationwide has also led the way in developing mix adjusted series.

The Nationwide has progressively improved its weighting methods. Adjustments for average prices per square foot were introduced in 1959 followed by a more extensive range of variables in 1977 (carried back to the first quarter of 1973 and shown in *Spon's house price data book*). The current mix adjusted series was introduced in 1983 and it includes not only the physical house characteristics but also a range of locational criteria; in that the Nationwide index differs from the other leading indices. The main criteria are set out below:

Agricultural areas.
Modern family housing, high incomes.
Older housing of intermediate status.
Poor quality, older terrace housing.
Better-off council estates.
Less well-off council estates.
Poorest council estates.
Multi-racial areas.
High status non-family areas.
Affluent suburban areas.
Better-off retirement areas.

The surrounding areas are also classified into seven broad groupings, e.g. inner metropolitan, rural areas and market towns, etc.

The full range of statistics was published as from the beginning of 1983; a run of figures for all dwellings carried back to the fourth quarter of 1973, nationally and by region, is available from Nationwide.

d) The Halifax Building Society

In 1977, the Halifax began supplying simple unadjusted average house price data for monthly publication in *The Times*. Calculated back to 1975, the data was made available by the Halifax with the explanatory notes to its new 1983 based index; they can also be found in *Spon's house price data book*. In April 1984, the Halifax introduced its standardised house price index, calculated retrospectively to 1983 but not linking with the earlier *Times* series. The Standardised House Price series was based on 12 different house price characteristics which, for interest, are set out below:

Purchase Price.
Economic Planning Region.
Type of property (e.g. detached house, terrace, flats, etc.).
Age of property in years.
Tenure.
Number of habitable rooms and bathrooms.
Number of separate toilets.
Central heating.
Number of garages and spaces.
Garden (yes or no).
Land area if greater than one acre.
Road charge liability.

Thus, although it can be seen that the index contains a large number of variables, all of which are important in determining the price of a house, it still leaves scope for substantial divergence within ostensibly uniform categories. That is not a criticism of the index, but more a criticism of the way in which the Halifax and other indices are some-times presented and used.

How accurate are the indices?

Two issues are involved: the speed with which the indices respond to changes in the market place and their ability to measure the scale of price movements. Even with the benefit of hindsight, there are no defini-tive statistics by which one can judge the accuracy of the data which was being produced contemporaneously. However, one can compare the extent to which the different indices told the same story, both nationally and regionally, and, albeit subjectively, see how the indices compared with the estimates of price movements made by those directly involved in the industry. Here the course of the price movements since 1988 are shown in the three leading mix adjusted series and they do show some interesting divergences. It is some relief that all three indices peaked in

Table 5.1 *Increase in house, prices, all dwellings UK*

	1985		1989		1991		1992
			% change on 1985		% change on 1989		% change on 1989
	Q1	Q3		Q4		Q4	
Halifax (1983 = 100)	112.2	227.3	+103	217.5	−4	199.5	−12
Nationwide (1983 Q1 = 100)	126.2	238.6	+89	205.2	−14	199.8	−16
BSA (1985 = 100)	94.7	208.0	+120	197.0	−5	182.0	−12

the same quarter (the third quarter of 1989); Table 5.1 shows the increase in house prices in the four year run up to that date, and the decline since.

Because of the magnitude of the increase in house prices between 1985 and 1989, the difference between the three indices does not look too large but, starting with each index as 100, the Nationwide index was 12% higher than the Halifax in 1985 and 6% lower in 1989. Between 1989 and 1991 the Nationwide showed a fall of 10 points more than the Halifax and the BSA. Indeed, a conversation with any national builder or estate agent would suggest that the Halifax and BSA figures were then understating the reality of the marketplace. During 1992 the series have converged to some degree and, although the cyclical movements still differ, it is interesting to see that the overall change in prices since 1983 is now virtually identical for the two building societies.

A complication in looking at any national index of house prices is the very considerable regional variation that has been experienced in recent years with northern house prices continuing to rise as those in the south have collapsed; only in 1992 did prices in the Northern region start falling. Anecdotal evidence suggested that house prices in the South East began to fall in the third quarter of 1988 (the double mortgage relief expired in August), and certainly by the fourth quarter. All three South East series examined in Table 5.2 (two of which are based on mortgage approvals) continued to record strong price growth in the fourth quarter of 1988. The Halifax index levelled out in the first quarter of 1989 and did not start to fall until the second quarter; the Nationwide did not even fall, quarter on quarter, until the third quarter and the BSA index until the final quarter of 1989.

Table 5.2 *House prices, all dwellings South East*

	Record high	(Period)	1992 Q4	% change
Halifax	259.9	(88 Q4)	181.4	−30
Nationwide	286.2	(89 Q2)	183.2	−36
BSA	218.0	(89 Q3)	167.0	−23

Eventually, both building societies appear to have produced a fall which is close to the price reductions which housebuilders admit to experiencing during the recession in the South East. One might then ask, why do the indices appear to have been slow to respond to the changes in the market place. One possibility is that the sellers of any product, one of whose virtues is financial gain, do all they can to sustain the list price. The reduction in the net selling price comes first through incentives which do not form part of the contract selling price, e.g. mortgage subsidies, free legal and moving fees, and the buying of the purchaser's own house at an inflated value. The scope for these inducements in the second-hand market is less but contents can be included at unrealistic values. The discrepancies between the indices are currently being studied by a DoE working party.

HOUSE PRICE EARNINGS RATIOS AND AFFORDABILITY

Why look at house price earnings ratios?

The house price earnings ratio is regarded as an important measure, perhaps the most important, of people's ability to buy a house. By linking house prices to earnings, it provides a convenient way of relating changing house prices to current purchasing power. Like many economic ratios which are adopted as fashionable yardsticks, the rationale for using this particular measure is rarely questioned. The use of the earnings ratio is predicated on the assumption that the marginal buyer will, in normal times, buy the most expensive house he can afford, i.e. what he can afford will depend upon what he can borrow, and what he can borrow and afford to finance will be a multiple of his earnings. A house price earnings ratio significantly above the historical average is taken as an indication that borrowers are extending their capacity to borrow, and therefore house prices 'look expensive'; conversely, a historically low house price earnings ratio is taken as suggesting house prices are low.

Within limited budgets, the concept of purchasing the most expensive product that can be afforded is not unique to housing (it exists in elements of the car industry for instance) but it is probably only a very small proportion of housebuyers since the Second World War that have consciously bought a house priced at significantly below their affordable limits; this will be particularly true of first time and first move purchasers.

It is sometimes assumed that only demand affects house prices and that in the long term there is little flexibility of supply. In other words, there is no long-term possibility of there being such a supply of houses (or more properly, the land on which one has to put them) that prices will grow more slowly than earnings. If it is argued that house prices will, in the long term, remain within a fixed range of house price earnings ratios then a house price linked to earnings means, of course, that house prices will rise relative to other output prices. Since there is no economic reason why the price of building materials will increase faster than the price of other manufactured goods, that would suggest that the price appreciation enjoyed by the holders of land will be correspondingly faster than the growth in incomes. To avoid that, there would need to be a secular decline in house price earnings ratios.

Which house price earnings ratio?

House price earnings ratios combine the errors implicit in measuring both house prices and incomes and therefore need treating with even more caution than the price series on which they are based. The reliability of house prices has so far been discussed in terms of their cyclical variation; there is also a problem with the reliability of the price series over long periods of time, for houses purchased today are certainly not directly comparable with those purchased 20 years ago. This is allowed for in the Nationwide moving weights, but not in the Halifax or DoE fixed weight systems. Long-term comparisons of house price earnings ratios must therefore be made with caution.

Unlike house price indices, the earnings measurements are not always drawn from the direct experience of the organisation producing the index. Some indices measure the earnings of the specific buyers of the houses that form the index; the alternative is to measure house prices against an index of national average earnings, only a small proportion of whom are actually buying houses at any one time. The first approach more closely mirrors the experience of the actual transactions that are taking place, but they can flatten out cyclical movements. For instance, as prices rise, fewer people can afford to buy and the transactions will gravitate towards those who are enjoying above average income growth; the house price earnings ratio reflects the financial influence of the diminishing pool of people who can afford to buy and will therefore rise more slowly than the rise in house prices might have suggested. It is for this reason that some indices use national average earnings as an indicator of what the community as a whole can afford. The disadvantage of the national earnings figures is that they can be removed from the reality of the marketplace as they include the earnings experience not only of the potential house buyer, but also those householders who have neither the intention nor the sustainable financial ability to engage in house purchase. The practical use, therefore, of the house price earnings ratio, must be treated with circumspection.

a) *The Department of the Environment/BSA Five Per Cent Sample*

Although the source material is the same, the DoE *Housing and Construction Statistics* sets out the price earnings ratios in a more convenient way than *Housing Finance*. Although the latter shows separately for the first time buyer and former owner occupiers, the average dwelling price and average income of borrowers, it leaves the reader to calculate the actual ratio. *Housing and Construction Statistics* (Table 1.13) sets out the ratios and also has a combined figure for all mortgages; it is, therefore, the preferred source for that series. Based on its five per cent sample of average house prices, and the recorded income of its borrowers, the DoE has published a house price income ratio since the first issue of *Housing and Construction*; this goes back on a quarterly basis to 1969 and annually to 1967.

The disadvantage of the five per cent sample series for house price earnings ratios is that it does not use mix adjusted prices. As it also uses the incomes of actual borrowers, the ratio is both lower and less prone to cyclical fluctuation than those Halifax and Nationwide series based on mix adjusted prices and national earnings.

Housing Finance does contain annual articles on house prices and earnings and its issue of August 1992 included an annual table of price income ratios from 1956 to 1992 (Q1), based on average prices at the mortgage approval stage, compared with national earnings.

Table 5.3 *House Price Earnings Ratio of the Halifax*

Index	Dec. 1988	Dec. 1992	% change
Standardised	4.76	3.31	−30
Not standardised			
First time buyers	3.71	3.58	−4
Former owner-occupiers	4.74	4.09	−14

Table 5.4 *Examples of Halifax regional house price earnings ratios*

Region	1st Q 1989	4th Q 1992	Change, %
Yorkshire	3.87	3.91	+1
South East	5.49	4.17	−24
UK	4.36	3.94	−10

b) The Halifax House Price Earnings Ratio

Dating back to the first quarter of 1983, the Halifax publishes a national price earnings ratio based on its standardised price index and national average earnings; because of the slower publication of the national earnings data this ratio tends to to run a couple of months behind the house price index. Additionally, it publishes price income ratios for first time buyers and former owner occupiers but these are based on the average price paid for a Halifax property (i.e., not standardised) and the income of the Halifax borrower; these two house price earnings ratios are produced contemporaneously with the house price index. Table 5.3 shows a striking difference between the ratio produced using the standardised price index and the national earnings figures, and those compiled from simple average prices and the incomes of the Halifax borrowers.

The Halifax has intermittently published regional house price ratios; although it no longer does so, the regional data continues to be prepared. The regional figures have been of particular interest in the 1989–92 housing recession when the experience of the north and south of the country have been so diverse. Unfortunately, although the official earnings series is available nationally on a monthly basis, regional earnings figures are only available on an annual basis. The Halifax regional price earnings ratios are therefore prepared from the average (unadjusted) price and recorded income of its borrowers, and presumably understate the fall in the ratio in the same way as shown in the Table 5.3. Examples are shown in Table 5.4.

c) The Nationwide House Price Earnings Ratio

The Nationwide has run a price earnings ratio since 1952 and this is based solely on the national average earnings data. The publication of the present mix adjusted series introduced a slight discontinuity and therefore care needs to be taken when using old bulletins. However, the Nationwide has recalculated the data back to the 1st quarter of 1974 and this now provides the longest consistent house price earnings series using mix adjusted prices. In fact, if a slight discontinuity in 1974 is ignored, the Nationwide house price earnings ratio can be extended back to 1954. No regional price earnings ratios are published.

d) Halifax v Nationwide

When the Halifax mix adjusted index began in 1983, the derived house price earnings ratio was a little above the long standing Nationwide series (3.43 against 3.21) but within a year the two had converged and between 1984 and the third quarter of 1987 the series were, to all intents and purposes, identical (see Table 5.5). Since then, however, the Halifax price/income ratio has run consistently ahead of the Nationwide and, in the fourth quarter of 1991, was 19% higher. This is the ratio to which analysts most frequently refer as an indication of the relative cheapness or dearness of the housing market.

The earnings denominator is common to both series; the discrepancy is totally explained by the different rates of growth in the two Societies' house price indices. Both societies argue eloquently why theirs is the true reflection of the market; *the advice here is not aimed at selecting one index rather than the other, but to warn against using any of the statistics in dogmatic fashion.*

Table 5.5 gives a comprehensive view of house price earnings ratios since 1969.

Table 5.5 *UK House price/Income ratios*

	1983/90	1984	1987	1989	1992		1992 Q4
	Low	*Q1*	*Q3*	*Q2 (Peak)*	*Q4*	*Below Peak*	*c.f. 1983/90 Low*
Halifax	3.43	3.47	3.87	5.00	3.37	−32%	−2
Nationwide	3.21	3.42	3.85	4.65	2.90	−38%	−10%

Table 5.6 House price earnings ratios since 1969

	1969				1970				1971			
	Q1	Q2	Q3	Q4	Q1	Q2	Q3	Q4	Q1	Q2	Q3	Q4
Halifax*												
Nationwide#	3.21	3.19	3.15	3.13	3.08	3.03	3.00	2.96	3.00	3.05	3.17	3.29
BSA†	2.68	2.66	2.62	2.59	2.60	2.57	2.59	2.57	2.52	2.51	2.59	2.64

	1972				1973				1974			
	Q1	Q2	Q3	Q4	Q1	Q2	Q3	Q4	Q1	Q2	Q3	Q4
Halifax*												
Nationwide#	3.46	3.66	4.00	4.07	4.17	4.21	4.23	4.34	4.41	4.10	3.89	3.66
BSA†	2.72	2.83	3.09	3.24	3.33	3.40	3.47	3.40	3.37	3.25	3.20	3.13

	1975				1976				1977			
	Q1	Q2	Q3	Q4	Q1	Q2	Q3	Q4	Q1	Q2	Q3	Q4
Halifax*												
Nationwide#	3.56	3.51	3.38	3.34	3.30	3.28	3.25	3.24	3.23	3.25	3.26	3.19
BSA†	3.06	2.98	2.90	2.82	2.73	2.74	2.76	2.72	2.63	2.60	2.63	2.65

	1978				1979				1980			
	Q1	Q2	Q3	Q4	Q1	Q2	Q3	Q4	Q1	Q2	Q3	Q4
Halifax*												
Nationwide#	3.24	3.25	3.47	3.57	3.65	3.75	3.83	3.89	3.84	3.74	3.60	3.49
BSA†	2.63	2.68	2.75	2.78	2.83	2.90	3.03	3.06	2.96	2.86	2.88	2.78

	1981				1982				1983			
	Q1	Q2	Q3	Q4	Q1	Q2	Q3	Q4	Q1	Q2	Q3	Q4
Halifax*	3.47	3.45	3.33	3.19		3.21	3.17	3.19	3.43	3.50	3.51	3.46
Nationwide#									3.21	3.29	3.33	3.34
BSA†	2.72	2.75	2.81	2.65		2.53	2.57	2.56	2.60	2.66	2.73	2.71

	1984				1985				1986			
	Q1	Q2	Q3	Q4	Q1	Q2	Q3	Q4	Q1	Q2	Q3	Q4
Halifax*	3.47	3.54	3.54	3.50	3.51	3.52	3.49	3.53	3.54	3.62	3.68	3.78
Nationwide#	3.42	3.52	3.50	3.54	3.55	3.55	3.52	3.54	3.52	3.55	3.61	3.68
BSA†	2.64	2.71	2.79	2.73	2.61	2.71	2.72	2.74	2.74	2.77	2.90	2.88

	1987				1988				1989			
	Q1	Q2	Q3	Q4	Q1	Q2	Q3	Q4	Q1	Q2	Q3	Q4
Halifax*	3.81	3.87	3.93	3.97	4.07	4.34	4.69	4.83	4.92	5.00	4.89	4.66
Nationwide#	3.78	3.85	3.90	3.80	3.84	4.02	4.37	4.45	4.56	4.56	4.61	4.38
BSA†	2.83	2.90	2.96	3.00	3.00	3.08	3.25	3.25	3.14	3.13	3.27	3.17

	1990				1991				1992			
	Q1	Q2	Q3	Q4	Q1	Q2	Q3	Q4	Q1	Q2	Q3	Q4
Halifax*	4.52	4.45	4.36	4.24	4.14	4.13	4.01	3.88	3.68	3.65	3.57	3.37
Nationwide#	4.18	4.01	3.82	3.58	3.50	3.50	3.43	3.29	3.14	3.14	3.10	2.90
BSA†	3.07	3.08	3.04	3.02	3.03	3.00	3.09	3.03	2.94	2.96	2.92	2.84

* Standardised all houses.
# Mix adjusted all houses.
† 5% sample all mortgage completions; there is a considerable variation in the income details recorded by different societies.

Affordability

The relationship of the house price to income is a very crude measure of affordability and it owes its prominence to its ready ease of calculation and the guidelines of the lending institution which typically advance a given multiple of salary. The other lending criteria, of course, is income related, i.e., what proportion of income is taken in servicing the repayment of the mortgage. The repayment ratio is dependent not only on the size of the loan (in turn determined by the house price) but the rate of interest; taxation interacts with those ratios in that a period of falling standard tax rates, such as occurred during the 1980s, produces greater disposable income, while the reduction in the real level of tax relief on mortgage interest works in the opposite direction. This is demonstrated in Tables 5.7 and 5.8.

The Nationwide used to produce an excellent table showing mortgage repayments as a percentage of net and gross income; the series dated back to 1973 but it was discontinued in 1990 as the Society ceased recording its borrowers' incomes. The article in *Housing Finance* in August 1993, cited above also contained an analysis, for first time buyers, of net repayments as a proportion of borrowers' incomes on an annual basis from 1969 to 1990 (see Table 5.9); these are updated and presented quarterly in the monthly *CML Market Briefing*. The ratio is derived

Table 5.7 *The reducing impact of tax relief*

		Rate at which which tax relief allowed	
Year	*Highest rate %*	*Standard/ basic rate %*	*Proportion of mortgage interest met by tax relief %*
1975–76	98	35	39
1978–79	98	33	34
1981–82	75	30	28
1984–85	60	30	28
1987–88	60	27	23
1988–90	40	25	21
1989–90	40	25	19
1990–91	40	25	18
1991–92	25	25	17

Source: Housing Finance, *No. 11, August 1991, p. 17.*

Table 5.8 *Tax relief and new mortgage loans*

			Percentage of loan qualifying for tax relief		
Year	Tax relief ceiling	Average new mortgage	Average mortgage	Twice average mortgage	Four times average mortgage
1975–76	£25 000	£7 890	100	100	79
1978–79	£25 000	£10 750	100	100	58
1981–82	£25 000	£14 990	100	83	42
1984–85	£30 000	£20 430	100	73	37
1987–88	£30 000	£28 590	100	52	26
1988–89	£30 000	£34 810	86	43	22
1989–90	£30 000	£38 780	77	39	19
1990–91	£30 000	£43 340	69	35	17
1991–92	£30 000	£45 000	67	33	17

Source: Housing Finance, *No. 11, August 1991, p. 18.*

Table 5.9 *First time buyers: initial mortgage repayments as a percentage of income*

1970	1971	1972	1973	1974	1975	1976	1977	1978	1979	1980	
11.4	11.6	12.3	15.5	16.6	14.7	13.7	13.2	11.7	15.9	18.9	

1981	1982	1983	1984	1985	1986	1987	1988	1989	1990	1991	1992
17.3	15.4	13.5	15.1	16.8	15.8	16.3	16.7	21.8	25.5	22.2	17.6

	19	89			19	90		
Q1	Q2	Q3	Q4	Q1	Q2	Q3	Q4	
20.4	21.2	22.1	23.8	24.7	25.3	26.4	24.7	

	19	91			19	92		1993
Q1	Q2	Q3	Q4	Q1	Q2	Q3	Q4	Q1
24.6	21.4	20.6	19.4	18.3	18.1	17.7	14.2	13.2

Source: Housing Finance, *August 1993, Table 6.*

from the DoE 5% sample series on house prices and earnings and must therefore be treated with the caution expressed earlier in the section on house prices. Nevertheless, viewed as a long term trend, the series provides a helpful guide to movements in affordability.

Table 5.10 *NHBC First time buyers ability to buy index*

Year	Q1	Q2	Q3	Q4
1975	71	77	82	84
1976	89	93	91	87
1977	88	90	96	101
1978	107	103	99	90
1979	88	86	79	63
1980	62	67	72	78
1981	82	89	88	91
1982	96	96	102	104
1983	104	98	93	94
1984	97	90	80	89
1985	77	72	79	81
1986	91	97	98	87
1987	83	83	88	88
1988	95	97	80	77
1989	65	51	42	46
1990	64	67	78	91
1991	89	91	95	101
1992	103	100	103	112

Source: NHBC Private House-Building Statistics 1992 (Quarter 4).

The highest ratio recorded on an annual basis during the whole of the 1970s was 16.6% while, in a number of years, including the beginning and end of the decade, the ratio had fallen below 12%.

Another useful indicator of affordability currently published on a regular basis is the National House-Building Council's 'First time buyers ability to buy index' (see Table 5.10) which takes account of the house price, the average deposit paid by first time buyers, the rate of interest and average earnings. The index is derived from a formula set out in *Forecasting Housing Booms and Slumps* by Hall and Richardson, Housing Research Foundation, 1979, and the series is available on a quarterly basis back to 1975. For the user, it has the added advantage of being published in full in each issue! The relationship between the index and activity in the housing market is discussed within the chapter on forecasting (Chapter 10).

Long term demand for housing

HOUSEHOLD FORMATION

It is difficult to think of any other product where there has been a closer long-term relationship than that which exists between one household and one dwelling unit. The rationale is straightforward: there is a strong drive on the part of family units to secure for themselves a house of their own. It is commonplace to talk of 'a right' to minimum housing standards and the individual's aspirations are therefore supported by a panoply of social, political and financial encouragements. However, once a family has acquired its primary accommodation, the cost of acquiring a second unit is normally prohibitive; the marginal benefit from that second home is very low in comparison with the first, and the financial and social framework is no longer favourable. Thus, although the quality of housing purchased increases with income, it is only a very small proportion of the population that occupies less than, or more than, a single dwelling unit.

Over the last century the total number of houses, or the housing stock as it is referred to statistically, has increased broadly in line with the

Table 6.1 *Households and dwellings: England and Wales*

Census date	Total dwellings	Households	Dwellings/ Households %
1911	7 691	7 943	96.8
1921	7 979	8 739	91.3
1931	9 400	10 233	91.9
1938	11 400	11 300	100.9
1951	12 530	13 259	94.5
1961	14 646	14 724	99.5
1971	17 024	**16 876	100.9
1981	18 995	18 334	103.6
1991	20 748†	20 131	103.1

† *December 1990* ** *Revised by DoE*
Source: to 1971 – Housing Policy, *DoE 1977, Part 1, p. 15;*
for 1981 and 1991 – Housing and Construction Statistics
1981–1991, Tables 9.3 and 9.10.

*Note: Since 1911 there have been changes of definition of both households
and dwellings; further background is given in* Housing Policy, *DoE
1977, Part 1, pp. 14–15 and* Household Projections England
*1989–2011. For instance, under the 1971 Census definitions, people
with a room of their own and catering separately, but sharing a sitting
room, were counted as separate households; under the 1981 Census
definitions, they were counted as one household.*

number of households (see Table 6.1). Reductions in the ratio between
dwellings and households have primarily been caused by the two world
wars.

The starting point for any assessment of the long term housing
requirement, therefore, is going to be the expected number of households. The total number of households, in turn, depends primarily
on the size of the adult population and the extent to which that population divides itself into discrete groupings or households. A complicating
factor is that, although the long term requirement for housing is determined
by the number of households, the number of households is, itself,
also influenced by the number of houses available. 'The interaction is
complicated, and it would be misleading to describe the quantitative
aspect of housing progress as if it were provision of dwellings to house an
independently determined total of households.' (DoE, *Housing Policy*,
1977, Part 1, p. 16).

Throughout the century, the number of households has grown faster
than the adult population which, until the 1960s, had in turn been
growing faster than the total population. (The average household size, or

Table 6.2 Population and households: England and Wales

Census date	Total population			Population aged 20 and over			Total households			
	000	Index	% increase	000	Index	% increase	000	Index	% increase	Size*
1911	36 071	100.0		21 683	100.0		7 943	100.0		4.54
1921	37 887	105.0	5.0	23 883	110.1	10.1	8 739	110.0	10.0	4.34
1931	39 952	110.8	5.5	26 998	124.5	13.0	10 233	128.8	17.1	3.90
1938e	41 215	114.3	3.2	28 670	132.2	6.2	11 300	142.3	10.4	3.64
1951	43 758	121.8	3.7	31 362	144.6	9.3	13 259	166.9	17.3	3.30
1961	46 105	127.8	5.4	32 321	149.1	3.1	14 724	185.4	11.0	3.13
1971	48 750	135.2	5.7	33 860	156.2	4.8	16 876	212.5	14.6	2.89
1981	49 634	137.6	1.8	35 414	163.3	4.6	18 334	230.8	8.6	2.71
1991e	50 903	141.1	2.4	37 960	175.1	7.2	20 131	253.4	9.8	2.53
2001f	52 526	145.6	3.2	38 767	178.8	2.1	21 775	274.1	8.2	2.41

Sources: Housing Policy, *DoE 1977, Part 1, p. 12;* Office of Population, Censuses and Surveys Population Projections (OPCS), *p. 2, No. 17, Appendix I.*
e estimate   f forecast   * Average household size

the relationship between population and household numbers, is normally referred to by the statisticians as the 'headship' rate.) This headship rate has fallen significantly through the century but really this is a derived answer and not an underlying explanation. Table 6.2 shows the consistently higher rates of increase in the number of households compared to the population increase.

In simple stages, the forecasting progression can therefore be seen as:

Total population;
of which adult population.
The headship rate or household size gives:
Number of households.
The ratio of houses to households gives:
Size of housing stock.

Population estimates

For the purposes of forecasting housing demand, the population forecasts can be taken as given. Those wishing to obtain a more detailed description of the forecasting principles behind the population projections will find a good general description in the introduction to the OPCS's *National Population Projections*. What is important for the construction analyst is to be aware where the sensitivities lie. There will be few who are preparing construction forecasts beyond ten years, and within that time span much of the population data can be considered fixed. As household formation is largely derived from the forecast of adult population, the number of births in the next ten years has no significant relevance (births normally follow a decision to form a household, although it is accepted that there are cases where they might be a preceding cause). The death rates over a ten year period determine the dissolution of households but, over that period, the rates are stable. However, where there is considerable fluctuation is in national migration; this is the main sensitivity in the short term population forecasts both nationally and regionally.

Table 6.3 shows the composition of the population increase over the last decade and that projected for the 1990s.

The migration component is a residual of much larger gross immigration and emigration totals; moreover, the migration numbers have been particularly erratic in recent years as historic net emigration turned into

Table 6.3 *Composition and sensitivity of population change: England and Wales*

| | 1981–91 | | 1991–2001 | |
	000	*Sensitivity**	*000*	*Sensitivity**
Population at start	49 634		50 903	
Births	6 642	5.2%	7 090	4.4%
Deaths	5 757	4.5%	5 637	3.5%
Natural increase	885		1 453	
Migration	384		170	
Net increase	1 269		1 623	

* *Effect of 1% change in births or deaths on % increase in net population.*

Source: 1989-based National Population Projections, *p. 39; the table also gives equivalent figures for Great Britain and the UK.*

Table 6.4 *Population change: England and Wales (000)*

| | 1981–1991 | | | 1991–2001 | | |
	1985-based	*1989-based*	*Change*	*1985-based*	*1989-based*	*Change*
Births	6670	6642	−28	7159	7090	−69
Deaths	5757	5757	−	5717	5637	−80
Natural increase	913	885	−28	1442	1453	+11
Migration	179	384	+205	4	170	+164
Total increase	1092	1269	+177	1446	1623	+175

Source: 1989-based National Population Projections, *p. 39;* 1985–2025 Population Projections, *p. 47.*

net immigration (at least for England and Wales). Opinion has been divided as to whether the recent trend is an aberration in an otherwise consistent pattern of population loss, or a long term reversal in the underlying trend. However, the broadly neutral assumption made, for instance, in the 1985-based projections, are now replaced by an assumption of continued net immigration, albeit at lower levels than in the 1980s. Table 6.4 compares the 1985- and the 1989-based projections and clearly demonstrates the scope for medium term error that lies within the migration assumptions.

A discussion of migration trends and assumptions can be found in the *1989-based National Population Projections* (pp. 8–10). In fact, the central assumption for the United Kingdom for future net migration remains zero. The net immigration of 170 000 projected for England and Wales in the 1990s is no more than the counterpart to the anticipated migration loss of 100 000 from Scotland and 70 000 from Northern Ireland.

Analysts forecasting long term trends in the housing market might be best served by forming their own view on net migration, while relying on the official forecast of the natural increase in the population.

Regional population projections

Regional population projections have been made for England since 1965, to assist in the wide range of local planning decisions. Whereas the national projections are prepared by the Government Actuary, the regional projections are made by the OPCS in consultation with the regional and local authorities. These locally derived projections are 'constrained to be consistent with the national projections for England prepared by the Government Actuary', a no doubt interesting political exercise. Disaggregated data will always exhibit greater fluctuations than the total but, by the standards of population statistics, there are some surprisingly high variations in the regional and local population figures. This is of more relevance to the housing market than are the aggregated figures; it is self-evident that a surplus of houses in city centres suffering population loss cannot be transferred to the outlying suburbs. Table 6.5 highlights the main regional movement during the 1980s and those expected in the 1990s.

Each of the broad regional divisions shown in Table 6.5 contains further divergence and this is brought out by the projections made at a county level. Table 6.6 shows the top ten and the bottom ten English counties ranked by size of percentage changes in population.

Household formation

In moving from changes in the population to changes in the number of households, the short term assumptions become much more variable and a wider range of possible answers becomes available. Discussion papers presented by housebuilders will tend to veer towards different assumptions than those prepared by county planners. Official estimates of the future number of households are published by the Department of the Environment, the latest being *Household Projections England 1989–2011* in August 1991. All tables contain the three base years 1989–1991,

Table 6.5 Population change England (000)

Region	1981 base	Change 1981–89				1989 base	Change 1989–2001			
		Nat*	Mig*	Total	% change		Nat*	Mig*	Total	% change
North	3 117	14	−58	−44	−1.4	3 073	43	−59	−16	−0.5
Yorks and Humberside	4 918	45	−23	22	0.4	4 940	130	−22	108	2.2
E. Midlands	3 853	56	90	146	3.8	3 999	129	134	263	6.6
E. Anglia	1 895	22	128	150	7.9	2 045	47	161	208	10.2
Gtr. London	6 806	175	−224	−49	−0.7	6 756	492	−415	77	1.1
Other S. East	10 205	153	270	423	4.1	10 628	377	302	679	6.4
South West	4 381	−9	280	271	6.2	4 652	44	377	422	9.1
W. Midlands	5 187	105	−75	29	0.6	5 216	198	−83	115	2.2
North West	6 459	61	−140	−79	−1.2	6 380	184	−191	−7	−0.1

* Nat: Natural increase. Mig: Increase from migration.

Source: 1989-based sub-national population projections for England. Population Trends No. 66, Winter 1991.

Table 6.6 *Largest population movements 1989–2001: change, %*

Cambridgeshire	14.4
Buckinghamshire	14.1
Dorset	12.2
Wiltshire	11.9
Northamptonshire	11.5
Somerset	11.4
Shropshire	10.4
Cornwall and Isles of Scilly	10.2
Devon	10.0
Berkshire	10.0
West Yorkshire*	1.5
Surrey	1.3
Greater London*	1.1
South Yorkshire*	−0.1
Greater Manchester*	−0.6
Durham	−1.8
Tyne and Wear*	−1.9
West Midlands*	−1.9
Cleveland	−4.1
Merseyside*	−6.5

* *Metropolitan county*

Source: *1989-based sub-national population projections for England.* Population Trends No. 66, *Winter 1991.*

followed by five yearly projections to the year 2011. The central estimates are by type of household and by region; more detailed breakdowns follow.

Household Projections had been produced at two-yearly intervals during the 1980s (known as the 1981-, 1983- and 1985-based Estimates of Numbers of Households) but for reasons of economy there was a four year gap before the 1989-based projections were produced in 1991; it is not clear when the next series will be available.

Household formation is defined as the increase (or occasionally decrease) in the absolute number of households between any period of years. The *Household Projections* booklet does not show household formation itself but this is simply a matter of deducting the totals in one year from another and then, if required, converting the figure to an annual rate. This is the basis for the more frequently produced tables which show the change in household formation.

Table 6.7 shows the central forecasts of the 1989-based projections which are compared with the 1985-based projections. Thus, the table not only shows the composition of the most recent official projection of the number of households over the next decade, but it also shows the extent to which the forecasts can change with the addition of a few years extra statistics, or with slight changes in the underlying assumptions.

Table 6.7 *Numbers of households: England (000)*

	1991		1996		2001	
Projection for	1985-based	1989-based	1985-based	1989-based	1985-based	1989-based
Household types						
Married couple	10 559	10 572	10 449	10 342	10 350	10 142
Lone parent	1 868	1 891	2 013	2 148	2 074	2 336
One person	5 099	5 093	5 691	5 756	6 184	6 354
Other	1 378	1 481	1 465	1 665	1 475	1 771
All households	18 903	19 036	19 617	19 910	20 083	20 603

These estimates of household numbers permit the calculation of incremental household formation.

	1986–91		1991–96		1996–2001	
5-year Period	1985-based	1989-based	1985-based	1989-based	1985-based	1989-based
Household formation	859	899	714	874	466	693
At an annual rate	172	180	143	175	93	139

Thus, in summary, the two sets of household formation forecasts can be compared as follows:

	Household formation projections, annual rate		
	1986–91	1991–96	1996–2000
1985-based estimates	172	143	93
1989-based estimates	180	175	139

Further analysis is produced in the form of regional household projections, by standard economic region, and household projections by age and sex. The implications of the regional tables are self-evident while the age classification is an important starting point in any understanding of the demand for first time buyer homes or sheltered housing. Beyond that, there is a very detailed breakdown by locality and type of household, taking the reader into the number of married couples in Solihull or one person households in Bromley, sensibly considered outside the normal day to day requirements of the construction analyst.

The title of the *Household Projections* booklet referred to England; previous projections had grouped England and Wales together and an irritating discontinuity has therefore been introduced. The projections for Wales are published by the Welsh Office. The Scottish household statistics have always been prepared separately and not always on a basis that has been consistent with those for England and Wales; their projections only go as far as 2001, rather than 2011 for England and Wales. For the totals only, i.e. with no breakdown for type of household, the Department of the Environment collates the figures for England, Wales, Scotland and Northern Ireland; these are provided for 1951 and then annually from 1961 to 2001. This table can be obtained on request; the figures for Wales are published (along with those for the English regions) in *Housing and Construction Statistics 1981–91*.

When working with other housing series where Great Britain is the normal geographic entity, it will be easier for analysts to use the household formation totals derived from the Department of the Environment summary; however, for any analysis of the underlying movements it will probably be more convenient to drop back to England only. The total number of households for Great Britain since 1951, with projections to 2001, are shown in Table 6.8.

Methodology

The household formation figures are net. They are the balance between new households being formed (marriage of individuals previously part

Table 6.8 *Number of households and household formation: Great Britain*

1951	1961	1966	1971	1976	1981	1986	1991	1996	2001	
14.6	16.3	17.4	18.6	19.3	20.19	21.11	22.14	23.12	23.87	A
	1.7	1.1	1.2	0.7	0.9	0.92	1.03	0.98	0.75	B
	170	220	240	140	180	184	206	196	150	C

A – No. of households, millions.
B – Household formation between periods shown, millions.
C – Houshold formation at annual rate between periods shown, thousands.

78

of another household being the commonest method) and existing households being dissolved (typically on the death or institutionalisation of
the surviving marriage partner). Understanding would be simplified
if the projections were based on that underlying logic. Unfortunately
for the construction analyst, the household projections are not compiled
by reference to gross movements in households as such, but are derived
directly from the population estimates. The projected number of households is derived by dividing each class of population estimate by the
projected number of people within a household, or the headship rate
referred to earlier. The official forecasts of headship rates in this country
are broadly derived from the extrapolation of past trends. This approach
provides stability to the forecasting relationships but does not easily
cope with changes in trends, e.g. will the unprecedented level of house
repossessions amongst first time buyers lead to young people delaying
the point at which they form an independent household? Nor does the
headship approach easily lend itself to a lay discussion of changes in
assumptions: can you readily evaluate the forecast reduction in the
headship rate from 2.50 in 1990 to 2.41 in 1996?

The forecasting sequence begins with (to use the latest projections)
the 1989-based forecasts of population by age and sex; based on this,
the Government Actuaries Department provides projections of the
numbers in each marital status. Finally, the household projections (for
England) are prepared by the Department of the Environment's Building Research Establishment by applying their projections of headship
rates. The projected headship rates are derived from past headship rates
using a curve fitting method, with a weighting system to give greater
importance to recent trends.

The central projections from *Household Projections England 1989–2011*
were shown earlier. Table 6.9 shows the summary regional analysis and
projections by sex and age. For a run of regional figures going back to
1981, these are most conveniently obtained from the annual *Housing and
Construction Statistics*, which has the added advantage of including the
data for Wales.

Table 6.10 shows the composition of household formation by age and
sex, derived from the 1989-based projections; earlier figures are not
available on a comparable basis. The 1981 and 1986 household numbers
have been revised, but the revisions for those earlier years have not been
published.

Table 6.9 *Number of households and household formation by region (000)*

	1981	1986	H.F.	1991	H.F.	1996	H.F.	2001	H.F.
North	1 149	1 177	28	1 210	33	1 238	28	1 259	21
increase, %			2.4		2.8		2.3		1.7
North West	2 369	2 417	48	2 493	76	2 562	69	2 615	53
increase, %			2.0		3.1		2.8		2.1
Yorks and Humberside	1 830	1 893	63	1 982	89	2 061	79	2 123	62
increase, %			2.9		4.7		4.0		3.0
E. Midlands	1 411	1 495	84	1 602	107	1 702	100	1 784	82
increase, %			6.0		7.2		6.2		4.8
W. Midlands	1 868	1 949	81	2 045	96	2 127	82	2 187	60
increase, %			4.3		4.9		4.0		2.8
East Anglia	701	760	59	821	61	883	62	935	52
increase, %			8.4		8.0		7.6		5.9
Gtr. London	2 649	2 721	72	2 774	53	2 848	74	2 896	48
increase, %			3.7		1.9		2.7		1.7
Other S. East	3 711	3 969	258	4 226	257	4 485	259	4 698	213
increase, %			7.0		6.5		6.1		4.7
South West	1 636	1 755	119	1 882	127	2 004	122	2 107	103
increase, %			7.3		7.2		6.5		5.1
England	17 325	18 137	812	19 036	899	19 910	874	20 603	693
increase, %			4.7		5.0		4.6		3.5
Wales	1 009	1 040	31	1 095	55	1 137	42	1 172	35
increase, %			3.1		5.3		3.8		3.1

H.F.: Household formation.

Source: Housing and Construction Statistics 1981–1991, *Table 9.10.*

Table 6.10 *Number of households and household formation by age and sex: England (000)*

	1991	1996	Increase	2001	Increase
Male 15–29	1 883	1 751	−132	1 563	−188
increase, %			−7.0		−9.3
Female 15–29	713	703	−10	651	−52
increase, %			−1.4		−7.4
Male 30–44	4 467	4 557	90	4 783	226
increase, %			2.0		5.0
Female 30–44	915	1 184	269	1 897	213
increase, %			29.4		18.0
Male 45–64	4 827	5 194	367	5 419	225
increase, %			7.6		4.3
Female 45–60	771	958	187	1 139	181
increase, %			24.3		18.9
Male 65+	2 833	2 928	95	2 998	70
increase, %			3.4		2.4
Female 60+	2 627	2 635	8	2 653	18
increase, %			0.3		0.7
Total	19 036	19 910	874	20 603	693
increase, %			4.6		3.5

Source: Household Projections England 1989–2011, *Table 4.*

An alternative approach to forecasting household formation

Population and headship rates provide the basis of the official forecasts of household formation; its overriding advantage as a forecasting system is that it remains the simplest way of dealing with the population statistics as they exist. For those analysts wanting a ready forecast of household formation, the official *Household Projections* will be the natural choice. However, the analyst should be aware that an alternative approach to household formation – and one which is conceptually easier to follow – has been attempted. This consists of making separate forecasts of each of the individual component of household formation (e.g. marriages). As a more formal paper put it, these dynamic models are: 'based on the philosophy that, to study household composition, it is only necessary to monitor individuals' changes of household affiliation. The primary aim is an understanding of the dynamics of household formation in terms of processes such as marriage, divorce and leaving the parental home.' (I.E. Corner, *Household demography and the effective demand for new housing*, Building Research Establishment, 1991).

Table 6.11 *Households formed and dissolved: Great Britain (000)*

	1967 Actual	1981 Forecast
Married new households	373	395
plus immigrant households	56	48
less emigrant households	−80	−60
less households dissolved by death	−150	−189
less elderly households dissolved in other ways*	−37	−46
Identified net increase	162	148
Other net household formation	−12	−18
Total increase	150	130

* *e.g. institutionalisation.*

Unfortunately, data for what might otherwise seem the more logical approach is not easily available; there is insufficient comprehensive information on flows between types of households, and regular forecasts of household formation using this dynamic approach are not available. However, the concepts involved, and some of the individual component calculations, regularly surface in discussions of long-term housing demand, and it is important to be familiar with them.

One of the early attempts to prepare a forecast of household formation through its component parts was set out by A.E. Holmans of the Department of the Environment in 1970 ('A forecast of effective demand for housing in Great Britain in the 1970's', A.E. Holmans, *Social Trends*, No. 1, 1970). What is of particular interest is to set out Holmans' estimates of household flows for his base year (1967); we also show (see Table 6.11) what was then his most distant forecast (1981).

The flows identified by Holmans in Table 6.11 were not exhaustive but were thought to be the most important for the purpose of analysing the demand for housing. For analysts wishing to gain an understanding of the concepts behind the household flows, and the magnitudes involved, Holmans' 1970 article remains an excellent introduction. His main categories are summarised below.

Married new households Published data with forecasts taken from the Government Actuary's projections. An allowance of 15 000 a year was made for the decline in married couples living as part of another household, i.e. a reduction in 'sharing'.

Households dissolved by death The household is not normally dissolved when the first married partner dies but on the death of the widow or widower. From Census information, estimates were made of the proportions of single and widowed men and women in a given age group that were householders and, from that, estimates were made of the number of households dissolved through death.

Elderly households dissolved in other ways These were approximate estimates only.

Emigrant and immigrant households in the base period were derived from the International Passenger Survey; this does not show the household status of migrants which had to be inferred from the analysis of age, sex and marital status. Forecasts assumed a continuation of past trends.

Holmans stressed in his 1970 article that 'The principal uncertainty about the increase in the number of households . . . relates to the number of single, widowed and divorced people who will live as separate one-person households.' It was noted that the largest part of the increase had been among older people and, at that point, there had been no marked increase in the number of young single people living as separate households. 'But this is an area where forecasting is extremely difficult, in that the effects of supply and demand are very much intermingled.' That cautionary note was justifiable; household formation by young single people grew rapidly over the next two decades.

The next attempt by the Department of the Environment to project

Table 6.12 *Households formed and dissolved: Great Britain (000)*

Net movements in	1971	1976	1981*	1986*
Marriages	334	286	320	337
Other new households	122	145	153	165
Divorce and separation	16	25	27	30
Emigration and immigration	−4	−8	−11	−11
Death and other dissolution of elderly households	−248	−271	−301	−328
Other households dissolved	−15	−17	−20	−25
Total	205	160	168	168

* *mid-points used for marriages.*

Source: Housing Policy, *Technical volume, Part I, 1977, Table III.7.*

the flow of households formed and dissolved was in the 1977 *Housing Policy* Green Paper (see Table 6.12). More information was now being shown relating to households being formed other than through marriage.

A warning

Since the exercise referred to earlier, the problems of forecasting gross household flows have become more complex. This is outlined by Alan Holmans in 'The 1977 Housing Policy Review in Retrospect', *Housing Statistics*, vol. 6, July 1991 (well worth reading just for its dispassionate analysis of past forecasting errors). In particular, Holmans concludes that the growth of unmarried cohabitation has been the factor which has most complicated the analysis of gross household flows.

> *. . . virtually nothing is known about the 'flow' aspects of cohabitation. How many cohabiting couples form per year? How many subsequently marry (or re-marry) de jure? How many cohabiting couples part? . . . These are all parts of the jig-saw of household formation and dissolution. At the time of writing there appears no immediate prospect of finding these pieces of puzzle.*

Table 6.13 is derived from the four official household projections made in the last ten years and clearly shows the very considerable change in individual categories, and the total expectation for household formation.

Table 6.13 is not intended as a criticism of the official forecasts, for they are based on the best available data, and set out their working

Table 6.13 *Household formation by household type: England (000)*

			1991–96				1996–2001	
	1981	*1983*	*1985*	*1989*	*1981*	*1983*	*1985*	*1989*
Forecast for:			*based projections*				*based projections*	
Married couple	159	78	−110	−230	57	−21	−99	−200
Lone parent	44	74	145	257	13	30	61	188
One person	336	384	592	663	237	298	493	598
Other	−18	9	87	184	−47	−31	10	106
Total	522	544	714	874	261	276	466	693
Annual rate	104	109	143	175	52	55	93	139

Table 6.14 *Single person households: England (000)*

		1991	1996	Increase	2001	Increase
Male	15–29	418	411	−7	370	−41
	30–44	522	748	226	947	199
	45–64	493	622	129	778	156
	over 65	588	663	75	736	73
Female	15–29	258	248	−10	221	−27
	30–44	245	366	121	465	99
	45–59	349	430	81	535	105
	over 60	2221	2267	46	2302	35
		5094	5755	661	6354	599

Source: Household Projections England 1989–2011, *Table 4.*

assumptions. Rather, the table is a reminder of the extent to which changes in social practices – divorce, young people living independently – can affect household formation in a relatively short period of time. Ten years ago, the accepted view was that, by the second half of the 1990s, household formation would fall from around 150 000 a year to little more than 50 000 a year. That 50 000 a year for England has been progressively raised and now stands at 140 000 a year.

In evaluating the current medium term forecasts, due regard must be given to the propensity for change in the underlying social assumptions. It can be seen from Table 6.13 that the greater part of the increase forecast for the 1990s is derived from single person households and, in view of their importance, Table 6.14 has been included to show the composition of the increase by sex and age.

It would seem appropriate to finish this section by returning to Table 6.1 shown to illustrate what could happen to the requirement for new dwellings in this decade if the official projection of the households is correct (see Table 6.15). This requires another assumption, that of the dwellings to households ratio. This fell slightly in the 1990s, as the rate of new building fell behind household formation. However, the ratio has been increasing at 0.75% per decade as rising incomes have facilitated a reduction in the number of households forced to share accommodation, while second homes have also increased in number. Applying the 0.75% increase to the 1991 ratio would take it just back to that recorded in 1981. On that basis, there would be a requirement for a net (i.e. ignoring replacement) additional 1.8–1.9 million houses.

If the number of households represents 'need', the ratio in the right

Table 6.15 *Households and dwellings England and Wales:*
a possibility for the year 2001

Census date	Total dwellings	Households	Dwellings/ Households %
1911	7 691	7 943	96.8
1921	7 979	8 739	91.3
1931	9 400	10 233	91.9
1938	11 400	11 300	100.9
1951	12 530	13 259	94.5
1961	14 646	14 724	99.5
1971	17 024	**16 876	100.9
1981	18 995	18 334	103.6
1991	20 748†	20 131	103.1
2001	22 559	21 775	103.6
Increase 1991–2001	1 811		

† *December 1990.*   ** *Revised by DoE.*

hand column of Table 6.15 represents the extent to which economic prosperity enables that need to be translated into effective demand. At this stage, we begin to be drawn outside the narrow confines of this book but it is important to indicate the sensitivity of the projection to incomes. Table 6.16 shows the net housing requirement on three different assumptions about the dwellings/households ratio: that it remains the same as in 1991, that it increases by the historical average of 0.75%, and that it recovers the amount lost in the 1980s and then increases by the 0.75% historic average. In the context of the changes they have taken place through the twentieth century, these are reasonably wide assumptions yet they produce a range of required new housing no more than plus or minus 10% from the central estimate.

Throughout this section, there has been an implicit assumption that each dwelling unit is occupied and therefore any given number of units can be used to balance any given number of households. In practice, there is a pool of vacant dwellings which exists at any one time, and which is not available. Vacancy rates can be obtained from Census data and from sample surveys. The *English House Condition Survey 1986* (see page 89) reported a vacancy rate of 4.5% of the housing stock. Some 41% were classed as long-term vacant, primarily because of their poor condition; short-term vacant would include houses empty between changes in ownership (not all moves are same day) and second homes.

Table 6.16 *Housing requirement by 2001 based on differing dwelling/household ratios*

Census date	Total dwellings	Households	Dwellings/ Households %
1991	20 748	20 131	103.1
2001 A	22 450	21 775	103.1
2001 B	22 559	21 775	103.6
2001 C	22 711	21 775	104.3
Net increase 1991–2001			
A	1 702		
B	1 811		
C	1 963		

In comparing the balance between housing stock and households to obtain a measure of the crude housing surplus or shortage, full allowance must be made for a permanent vacancy rate. If a 5% vacancy rate is taken, then there remains a modest absolute shortage of housing stock as compared to the number of households.

REPLACEMENT DEMAND

Can replacement demand be measured?

The demand for new housing is a function of the change in the required size of the housing stock which, as indicated earlier, flows from household formation, modified by the ability to reduce sharing and to acquire second homes. There is also a requirement to provide for replacement of the existing stock. Replacement demand for housing is a simple concept but one which is almost impossible to quantify in any helpful way. Digress for a moment to the car industry, which has an average annual scrappage rate of around 6% of the total stock (SMMT research): halve the replacement demand for a few years and it follows that a significant backlog of demand will accumulate. There will always be arguments as to how long a car can be made to last, and the influence of income levels, but at least motor industry analysts would be able to have a meaningful discussion on the concept and quantification of replacement demand.

What are the equivalent figures for the housing market; how long do houses last? And, most important, how postponable is the decision to scrap a house (i.e. physically demolish it or leave it permanently empty).

Unfortunately, the range of data available is capable of almost any interpretation. The problem stems from the size of the housing stock in relation to the annual rate of new building. The annual average rate of new build during the 1980s was 200 000, or only 0.9% of the current stock of dwellings. In other words, to replace the housing stock on a 100 year cycle would account for the whole of the 1980s building programme; even a 200 year replacement rate would require 110 000 new houses a year. In reality, there appears to be nothing like that rate of attrition in the housing stock taking place. The Department of the Environment produces statistics showing the composition of changes in the housing stock and these are summarised in Table 6.17.

The 'other' gains to the housing stock comprise the conversion of larger units into a number of smaller dwellings, typically large houses into flats. The losses to the housing stock have been remarkably small. Slum clearance is down to almost nominal amounts compared with twenty years ago when around 100 000 dwellings a year were taken out. Other losses include fire, damage, obsolescence, change to other use, and demolition to make way for other development.

For those who draw comfort from statistics, there are plenty to play with (or statistically model). A breakdown of the housing stock by age is available from the Census and is updated between Census dates by the

Table 6.17 *Housing gains and losses: Great Britain (1982–91)*

	000	*Annual rate*
Housing stock December 1981	21 081	
Gains 1982–1991		
New construction	2 015	202
Other	162	16
Losses 1982–1991		
Slum clearance	(104)	(10)
Other	(107)	(11)
Net gain	1 965	196
Housing stock December 1991	23 046	

Source: Housing and Construction Statistics 1981–1991, *Table 9.2.*

Table 6.18 *Breakdown of housing stock by age (000)*

	Pre-1890	1891–1918	1919–1944	Post-1945	Total Stock
December 1972	3 702	2 973	4 488	8 017	19 180
April 1981	3 332	2 913	4 442	10 268	20 955
Change	−370	−60	−46	2 251	1 775
December 1991	3 226	2 904	4 379	12 537	23 046
Change	−106	−9	−63	2 269	2 091

Note: a pre-1871 division is available for areas other than Wales.

Source: Housing and Construction Statistics 1981–91, *Table 9.5; (original table shown in percentage form)*.

Department of the Environment; this shows surprisingly little reduction in the nineteenth century housing stock (see Table 6.18).

A dwelling unit will be taken out of the stock when it has no useful economic existence. This is substantially a matter of what can be afforded and houses are passed down the income chain in exactly the same way as cars. Public authorities may interpose their decisions for reasons of health and structural safety but even here there are no absolute standards, only those which increasing prosperity makes acceptable. The most informative physical measure of the condition of the housing stock is the *English House Condition Survey*, prepared every five years since 1971. The last survey was made in 1986 and the results published in 1988; a 1991 Survey has been made but the results (due for publication in September 1993) were not available at the time of writing.

The condition of the housing stock

Substantially greater detail is available on the structure of the housing stock in the *English House Condition Survey*, although there are a reduced number of age ranges (1919, 1944 and 1964 being the break dates). However, within these analyses, each age range is analysed by tenure and for each tenure by type of flats. Using these categories it would be possible to make more focused estimates of the requirement for replacement. Thus, it would be assumed that the post-war private stock would have a minimal replacement requirement within any normal forecasting time horizon, whereas higher figures would be assumed for, say, pre-1890 stock or post-war high rise local authority dwellings.

The housing stock was also assessed on the level of amenities (e.g. with bathrooms) and on the amount of repair work required. A category of 'Unfit Dwellings' (as most recently defined by the 1985 Housing Act) is recorded. Some 4.7 million dwellings were reported as defective in some respect but only 0.9 million (or 5% of the stock) were reported to be unfit for occupation. This 0.9 million houses represents the pool from which demolition may be drawn but there is no indication what proportion may be modernised and what proportion abandoned. To some extent, the poor quality of some of the older housing stock is already taken into account in the vacancy rate. Whereas the vacancy rate for the total *English House Condition Survey* was 4.3%, it was only 2.6% in the post-1945 housing stock but as high as 8.6% in the pre-1919 housing stock. This suggests that, although still available, these older units were harder to sell or let, and more frequently taken out of occupation for major refurbishment.

One of the problems in using the *English House Condition Survey* to determine long term trends in improvement or disrepair is that most tables compare only the last two survey periods; changes in definitions make the use of earlier volumes suspect. However, a graph displaying the three main criteria over time, on a constant basis, is provided and the hard data is reproduced in Table 6.19.

An earlier run of figures, though not directly comparable, taken from *Housing Policy 1977* indicates the progress that has been made since the war in improving the quality of the housing stock (see Table 6.20).

Returning to the *English House Condition Survey*, it is clear that the progress which has been made is in the installation of basic amenities (e.g. W.C., bath), primarily in the pre-1919 housing stock; three-

Table 6.19 *Dwelling lacking basic amenities, dwellings unfit, dwellings in serious disrepair (000)*

	1971	1976	1981	1986
Unfit dwellings	1216	1162	1116	1053
Lacking amenities	2815	1531	863	543
Serious disrepair	1104	1097	1178	1113

Note: these figures should not be totalled as dwellings may fall in more than one category.

Source: English House Condition Survey, *p. 74, 1986.*

Table 6.20 *Number of households occupying dwellings and that are unfit and/or lacking one or more basic amenities: England and Wales (000)*

1951	1961	1971	1976
7500	4700	2850	1650

Source: Housing Policy 1977, *p. 142.*

Table 6.21 *Demolition of dwellings: Great Britain (000)*

1971–75	1976–80	1981–85	1986–90
384	208*	86*	35*

*years to following March.

Source: Housing and Construction Statistics 1981–1991.

quarters of the change resulted from the provision of amenities, and the remainder from demolition or conversion to other use. However, the numbers of dwellings unfit or in serious disrepair has shown little change. Reductions in the pre-1919 numbers have come largely from demolition but the passage of time has added other dwellings; the number of inter-war dwellings in serious disrepair has, for instance, increased. For comparison, the numbers of dwellings demolished is shown in Table 6.21.

Contract types and contractual issues

The operational differences between various types of contract, and contract law, are outside the scope of this book; they have their own body of literature. Notwithstanding a little learning is supposed to be a dangerous thing, it is important for the construction analyst to acquire a layman's knowledge of the types of contracts typical within the construction industry together with some of the more financially sensitive sub-sections. This section discusses the main classes of contracts which the analyst will encounter, and then moves on to discuss specific contractual issues that will (or should) continually recur in any connection with a construction company (e.g. claims, bonding, etc.).

THE CONTRACT

Contracts are frequently referred to by one of their characteristics (e.g. management, fixed price) as if that were all to it. This can lead to misconceptions for contract definitions cannot always be self-contained; types of contract overlap and the names chosen by individual contractors may reflect their marketing philosophy rather than underlying commercial differences. The criteria most commonly used to categorise a contract reflect either the extent of the contractor's involvement in the construction process (construction only, design, management, etc.) or the method of remuneration (e.g. fixed price, escalation clauses, cost plus); these are sometimes presented as if mutually exclusive. If these characteristics are termed the 'involvement' criteria and the 'payment' criteria, the reader will readily appreciate that payment methods can apply to any level of involvement. Confusion arises because a contract may be referred to as fixed price or management as if no other criteria applied.

The principal contract descriptions are described here, listed either by involvement or payment. One final statement of the obvious: the structure of the contract matters less than the competence of the contractor and the design team in executing the work. The legal format tends to be less important in a successful contract than one in which there are problems, or where financial returns have been pared to the bone.

Involvement criteria

The employer, the contractor's client, initiates the construction process: he wants a structure – an office, factory, house extension or power station – and turns to a contractor to supply it. At that point, the employer has to decide who will provide the variety of functions necessary for the completion of any successful project: design, organisation, supervision and construction. The contractor is, obviously, involved in the last of these stages but it is possible for the contractor to participate in any or all of the other aspects of the project, and it is this depth of involvement that determines the type of contract which can be awarded.

Construction only

Construction only can be regarded as the traditional contract where the employer assumes responsibility for the design. The client, with the assistance of professional support (assuming anything other than a small contract), will specify the structure he wants and appoint a contractor to carry out the work. With the contract will go the conditions, the bills of quantities and the drawings which together set out the work to be done under the contract. The quantity surveyor measures the amount of work and materials necessary, sets them out in the bills of quantities and will frequently be involved in measuring the value of later changes in the work. The architect (for building work) and the engineer (for civil engineering) will provide the plans and drawings and will supervise the contract on behalf of the client.

Open tender

The method of choosing the contractor gives rise to different contract terminology. He may use an open tender system, advertising for competitive bids, normally choosing the lowest price. This is an approach more commonly found in the public sector where value for money is sometimes equated with the lowest price.

Selective and negotiated tender

Alternatives to the open tender are various forms of selective tender where the client (or his professional adviser) chooses a specific number of contractors and invites them to tender. The selection may be on the basis of past relationships, recommendations, or complicated pre-selection procedures to demonstrate competence to bid for large contracts. Public bodies usually maintain formal tender lists of contractors and one element of a contractor's goodwill is its presence on tender lists which other contractors may not have been able to penetrate. The selective tendering may be with as few as two or three contractors, and some relationships are so regular that individual contracts are negotiated with one contractor only. Alternatively, in repetitive work, the first superstore, garage, cinema or whatever, may be competitively tendered with the follow-up work being negotiated on the basis of the costs established in the first contract.

Once we move away from construction only contracts, we find that, in one form or other, the contractor has moved backwards (perhaps 'earlier' would be a less ambiguous word) to provide all or part of the design specification and management function. These contracts rest on the premise that the contractor's expertise in executing work should be harnessed at as early a stage in the specification process as possible: the arguments between using the men of theory or the men of practice will never stop, only ebb and flow. From now on, the contract terminology begins to overlap and confuse, but descriptions of the two main categories of contract follow.

Design and build

The logical conclusion to the contractor becoming involved in the specification process is for him to provide the design function in its entirety, working to the client's end requirements. The design and build contract eventually merges into the 'turnkey contract', with the latter taken to include every aspect of bringing the project into an operational status, leaving the client only to 'turn the key'. Thus, if the project was a hospital, the contractor would not only provide the design and carry out the construction but procure all the medical equipment and maybe even recruit the medical staff, stopping only at taking an arm and a leg from the client. 'Package deal' is also used to describe this type of contract.

Management contracts

With management contracts, the contractor is integrating backwards organisationally, managing the whole contract alongside the client, and undertaking supervisory functions otherwise carried out by the architect or engineer. As with design and build, the contractor is called in at an early stage; indeed, he may also be contributing to the management of the design process. It is a form of contract used for very large projects and the management contractor does little or no construction work himself but organises those who do.

> *It became a recognisable type of contract in the late 1970's and developed widely during the 1980's, when it was seen as the answer to the antagonistic and claim-conscious environment and the long mobilisation period associated with the traditional building contract. It offers the Employer the opportunity to harness the expertise and skills of an experienced Contractor and to ensure that the Contractor identifies solely to the Employer's objectives.* (George Capon, *Guide to the construction industry*, 1990)

The management contractor is usually paid the prime cost (his expense in providing the organisational function) plus a fee or bonus related to the amount by which the contract costs less than the target amount. Like the traditional construction only contract, the cost and fee may be determined by tender or negotiation. However, because the contractor's role is confined to mangement, he does not bear the normal contract risks and the margins on this type of work are correspondingly low.

Payment criteria

To quote from *Keating on building contracts* (5th edition, London, 1991): 'The manner of payment can be arranged in a variety of ways and it is impossible to attempt any exhaustive classification.' Undaunted, the description that follows is intended to cover the broad payment criteria most frequently used. These are also the criteria which can make or break the profit and loss account.

Fixed price or lump sum

Meaning exactly what it says, under a fixed price contract the work will be delivered at the pre-agreed price and the two contracting risks (that the work turns out to be more time consuming, or the materials and labour more expensive) are born by the contractor. Alternatively expressed, the contractor agrees to provide everything specified in the contract for a 'lump sum'. Fixed pricing is at its clearest when relating to the totality of a conventional construction only contract. For this reason, work of this type is often referred to as fixed price contracting, although the implication that other forms of contract do not have a fixed price element is misleading, for there are, of course, fixed price elements in every contract. Known originally as 'contracting in gross', the fixed price contract for the whole of a project (as opposed to a single trade) came into prominence in the early nineteenth century, as government committees questioned the cost overruns on public buildings (M.H. Port, 'The Office of Works and Building Contracts in Early Nineteenth-Century England', *Economic History Review*, 1967, pp. 94–110). The fixed price contract remained the standard before the Second World War, but inflationary times since then have led the contracting industry to seek what protection it can obtain on the longer term contracts. The fixed price remains the norm for all small works due to its simplicity and, for instance in the domestic market, the non-professional status of the customer; and for contracts of short duration, there is no necessity for the contractor to seek protection against rising costs. In the public sector, attempts to control inflationary pressures led to the insistence on all contracts under two years duration being let on a fixed price but pressure from the industry following the inflationary losses in the early 1970s led this to be reduced to one year.

Fluctuation clauses

The object of fluctuation, or escalation, clauses is to give some measure of protection to the contractor against rising costs, the extent of which would not have been easy to predict at the time of tendering; obviously, the longer the contract, the more likelihood there is that the contract will contain fluctuation clauses. This is not solely for the benefit of the contractor for if the client insists on a long term fixed price contract, he will have to pay the additional risk premium that contractors must inevitably build into their price. Escalation (or fluctuation) clauses do not offer total protection to the contractor: however widely they appear to be drawn up, there are two significant areas which are not protected. Firstly, the fluctuation clauses relate to site costs and do not cover the not inconsiderable head office and administrative costs that contractors incur. Secondly, the inflation adjustments relate to 'official' indices of labour and materials and to nationally negotiated wage rates. Thus, if buoyant market conditions drive rates above the official levels (this will largely be confined to labour and sub-contractors), then the 'excess' payments are not recoverable by the contractor; the same also applies to bonus payments.

The corollary is that in weaker market conditions the contractor may benefit by applying an escalation adjustment based on official indices, yet purchasing at discounted rates. In a recession, there is a tendency to maintain materials list prices and compete via discounts but it is the list price that is used in the preparation of the indices.

The judgement of future inflation rates and the extent to which the contractor seeks to change the weighting between his fixed price and his escalation contracts can be of crucial import in determining profitability.

Measurement

Payment at agreed rates for measured quantities of work was a traditional method of remuneration on contracts let to single or groups of craftsmen, and its importance lessened with the emergence of general contractors who could undertake whole works. Today, such payment systems may variously be known as time and fee, time and material, measure and value. Rather than price a contract in its totality, the contractor is presented with detailed lists of all work to be done together with approximate quantities. Each quantity is separately priced by the contractor and as the work proceeds, he will be paid according to his quoted rate and the quantities measured, with the quantity surveyor playing a crucial role on both sides. Although two estimates may be similar in total, the constituent elements may vary considerably. Part of the bidding skill is to have the more favourable unit prices on those items where the contractor thinks that quantities will exceed those originally estimated, or to have the higher prices for the earlier work to maximise the inflow of cash.

Cost plus/Reimbursement

The contractor's actual costs are reimbursed as incurred, sometimes directly to the supplier. Cost plus is particularly useful where it is difficult to specify the contract in detail, perhaps because of site problems or an unusual structure, or when speed necessitates beginning the project before it has been fully designed. One of the commonest areas to find cost plus contracting is the defence industry. The fee basis may be a percentage of the contract value, or a fixed fee, or a payment relating to savings against agreed target costs. The fee element may either be competitively tendered or negotiated.

CONTRACT ISSUES

Sub-contracting

The main contractor has the contractual relationship with the employer; the sub-contractor's relationship is with the main contractor. On large contracts with a high degree of specialist content, some sub-contractors may be chosen directly by the client, possibly even before the main contract has been awarded; these are referred to as 'nominated' sub-contractors. The sub-contractor may even have responsibility for design content, which can cause problems for the main contractors. The recession, with its consequent increase in financial failures – of employers, contractors and sub-contractors – has sharpened the focus on the precise relationships between the three groups. Is the sub-contractor entitled to be paid by the main contractor even though the main contractor has not been paid by the client? If a sub-contractor nominated by the client fails, what is the responsibility of the main contractor? These are complicated matters which are currently the subject of intense debate within the industry; suffice it to say here that the analyst should be aware that the problem exists.

Variations and claims

However familiar the construction techniques, every contract is a one-off sited on its own individual location. To state that problems are inevitable is perhaps to misuse the word 'problem'; let us say that the outcome is not always the same as the intention. To the extent that the contractor underestimates the resources that he needs to complete the contract, that is the contractor's problem and he has no recourse to the client. However, where the work to be done under the contract proves different from that originally specified, then the contractor will normally have redress under the terms to the contract. We have now entered one of the most contentious areas of the construction industry.

Variations

Variations and claims are sometimes confused by the outsider, possibly because they both cover changes from the original contract, and the former may also lead to the latter. The principle of the client wishing to vary his original specifications as the contract proceeds is well recognised and provision for that eventuality is contained in standard clauses within every contract. Variations is a term used within the construction industry in a technical sense to describe changes specifically requested by the client; it is not used in a lay sense to mean any event which turns out different than originally expected. Unfortunately, according to *Keating*, 'There is no generally accepted definition of extra work'.

As examples of variations: the client may want the doors varnished not painted; bigger trains to go through the tunnel; more powerful air conditioning in the office; deeper foundations. As well as changes to original specifications, the client may also request additional work: 'while you are here could you change the washer, repair the roof...'. All these are 'variations' from the original contract. Providing that variations have been properly authorised by the architect or consulting engineer, they should be certified and paid in the normal contractual manner.

Insofar as the original contract would have been competitively priced, the subsequently negotiated price for variations, when the client cannot call on alternative prices, can be a more profitable outcome for the contractor. Although the contract may stipulate the manner in which variations should be priced, to limit the contractor's charging flexibility, it is not unknown for contractors to bid very competitively on contracts where they believe that there are likely to be an above average proportion of variation orders. To return to our original distinction, a variation only becomes a claim when it is not agreed; more on this follows.

Claims

Again, *Keating* warns on the problems of definition: 'No exact meaning can be given to the term.' In the course of the contract, the contractor may be required to carry out work of a nature or quantity different to that originally tendered for, not because the client has asked him to vary it, but for reasons outside his control. Some reasons 'outside his control' are the contractor's own risk, adverse weather being a prime example. Others will be matters which a reasonably experienced contractor could not have been expected to foresee and for which the contractor would be able to 'claim' reimbursement against the client. It is an underlying principle of the construction contract that the contractor is not required to price for a risk he cannot see. This type of claim is more frequently found in civil engineering contracts because of the adverse physical conditions typically experienced. For example, the tender documents provided to contractors bidding for a tunnel may have specified the ground conditions as 90% clay and 10% rock; if the reality is that there is 20% rock the contractor will incur substantial additional costs for which he may be entitled to submit a claim.

Another category of claims would cover the client or other parties impeding the contractor's ability to proceed with the contract in the most expedient, and therefore profitable, manner. This would include restrictions on access to the site or delays in providing working drawings. These would generally be claims for consequential delay. Because, by definition, the changes to the contract are not agreed in advance, they are subject to considerable 'discussion' between both sides as to:

the responsibility for the variation or the event leading to the claim
(did the contractor partly contribute, could it reasonably have been foreseen?);
the physical measurement of the additional work or consequential delay;
what is a reasonable cost for the additional work;
do the contract conditions provide for the contractor to recoup all of that cost (especially relevant on some international contracts)?

To the extent that the client does not agree the contractor's figures for variations and claims, the contractor can either accept the position or, negotiation failing, proceed through arbitration or litigation. The accounting treatment of claims is discussed in Chapter 8.

Liquidated damages

This is a term the outsider will come across without ever being aware what is liquid about the damages. In the contractual sense it is a sum of money (or the damages) which is fixed and agreed (unliquidated damages is an ordinary claim for damages). Liquidated damages are paid in the event that the contract is completed after the agreed date (subject to extensions of time given by the employer under the terms of the contract) and it should represent the employer's genuine estimate of damage suffered by late completion; they are not to be regarded as a penalty (not enforceable under English contract law).

Bonding

Bonds are a financial guarantee supplied by a third party to cover the performance of the contractor, hence the expression 'performance bonds'. There are also tender bonds, provided to ensure that the contractor does not withdraw his tender within a stipulated period but these are normally found only in overseas contracts and have no particular analytic concern. Insofar as the performance bond is merely an independent guarantee of what the contractor was already committed to under the terms of his contract, they also should have only limited concern to the analyst. However, there are two aspects that must be mentioned. Firstly, if a contractor's overall financial position is becoming stretched, he will find it more difficult to obtain bonding and this, in turn, will impinge on his ability to secure work, and the cost of so doing. This is one of the ways in which a weak balance sheet can put pressure on trading.

The other aspect of bonding which must be viewed with great concern is the granting of 'on demand' bonds. The normal sequence of events if a client wishes to obtain damages from a contractor is to take legal action. If the client is successful, the contractor pays; if he defaults on the payment then the bond is called. The existence of a conventional performance bond puts no pressure on the contractor; it is there solely as protection of the last resort for the client. With the on demand bond, the client can call in the bond as he wishes. The guarantor (bank or insurance company) will pay the amount of the bond on the demand of the client, and then seek recourse from the contractor under the terms of the bonding agreement. The calling of the bond may have no commercial justification but the onus is then on the contractor to take the client to court and prove that the bond has been unfairly called. These on demand bonds are mainly found on international contracts, particularly in the Middle East, and some British contractors consistently refuse to tender for work under that type of bonding arrangement. It is always interesting to look at the contingent liabilities notes in contractors' accounts to see what is stated; it was rare in the 1970s for the contractors who had given on demand bonds to make mention of them. Where contractors are part of larger groupings, parent company guarantees may be required as well or instead of bonds.

Accounting issues

This chapter is not written as a mini-manual of construction accounting but as a layman's guide to the issues which are relevant to the construction industry in particular, rather than the corporate sector in general. For a detailed analysis, readers should refer to The Institute of Chartered Accountants' *Guide to the construction industry* by George Capon. What are considered to be the 'relevant issues' can vary with the current fashion and with the stage in the economic cycle; however, while other issues may come and go, the valuation and treatment of long-term work in progress will always remain at the heart of construction industry accounting. Work in progress valuation produces, arguably, a greater degree of accounting subjectivity than in any other industry. The other group of issues, which tends to come to the fore in times of recession, derives from the industry's role as a developer, both of commercial property and of speculative housing; here we think of land valuation when market prices are moving sharply, capitalised interest, the valuation of investment properties, and off balance sheet financing. As better times arrive, attention may more readily focus on the utilisation of earlier provisions and the disclosure of profits on the sale of land.

THE CONTRACTOR

The valuation of work in progress

In the major part of commercial activity – most manufacturing, distribution, services – the valuation of stock and work in progress (if there is any of the latter) is relatively simple. It is included in the balance sheet at cost unless it is thought to have lost value, in which event it is included at that lower net realisable value. For most industrial companies, the larger part of the stock purchased in a given year is also sold in that same year and the profit or loss can be ascertained directly from the totality of the transactions within the accounting period. Where stock or work in progress does remain at the year end, no profit contribution is taken on that stock.

Smaller construction companies – jobbing contractors, small subcontractors – will relate to the above description in that the larger part of their contract work will have been started and completed within the same year; any uncompleted work at the end of the financial year is valued at cost and no attempt is made to measure the profit element of work carried over for completion in the following financial year. Indeed, before the introduction of the first Accounting Standard on Stocks and Long Term Contracts in 1975 (SSAP9), that was a permitted treatment for any contract, of whatever duration. (Wimpey, for example, valued all contract work in progress at cost.) Thus, a long term contract might produce no profit until physically completed or even until that wonderful expression 'financial completion', when all the negotiations had been settled. In that way, a five year contract may not produce its profit contribution into the statutory accounts for several years and it was not unknown for contractors to keep open the final account on minor matters so that profit could be held back until required. Profit recorded in any one financial year might bear no relationship to the turnover of that year and, therefore, the profit margin calculated on the basis of one year's figures can be misleading. When analysing individual operating subsidiaries, this warning becomes all the more relevant.

This ultra-conservative accounting policy was deemed not to give a true and fair view of the state of the profit and loss account and, accordingly, SSAP9 required contractors to include, in their valuation of long term contracts, a proportion of the estimated profit on the contract, where the outcome of the contract could be foreseen with reasonable certainty. Most contractors moved cautiously towards the new rules and the changes to their published figures, if disclosed, were not large.

Wimpey, perhaps the most conservative of all because it treated housing sites as long term contracts, reported a 10% uplift on its adjusted 1975 pre-tax profits. Laing stated that its UK construction profits for 1976 were 16% higher on the new basis than the old although, of more interest in disclosing the swings that are possible, it also stated that its 1976 overseas construction loss of £137 000 would have been a profit of £1 021 000 under the previous accounting rules.

SSAP9 was revised in 1988 and it is to that standard that any subsequent comments refer. Although SSAP9 has led to a closer relationship between the work done and the profit recorded, there are factors which still contribute to a bunching of profit towards the latter part of the contract. Contractors cannot reasonably foresee the profit at the early stage of the contract and therefore no contribution is likely to be included; the nearer the contract comes to completion, the more probable it is that the profit can be foreseen with reasonable certainty.

Provision for losses

Accentuating the contration of profit towards the end of the contract is the requirement to provide for losses as soon as they are anticipated. Frequently these losses are theoretical (or 'accounting losses') in that if the nature of the contract changes, there may still be reasonable expectation that higher costs will be reimbursed as claims are submitted. However, until there is agreement on claims, the Accounting Standard requires those losses to be recognised; this is a particular feature of civil engineering. The corollary of those early accounting losses is that at the latter stages of the contract, or even after its completion, the claims negotiations are concluded and additional sums are agreed and credited to the profit and loss account.

The comments made so far relate to the incidence of profit at a site level, but below this is a not insignificant group or divisional overhead which may be allocated on a site basis. This overhead charge does not necessarily change with the flow of new work. If, for instance, there is a significant increase in new work which does not make a direct site contribution to profit, it may make an indirect contribution because it absorbs a proportion of the group overhead; this leaves a smaller overhead to be charged against contracts which are contributing profits. Thus, even though a finance director may be stating that profit is not taken until the later stage of a contract, the impact on overhead recovery may lead the new work to make an effective contribution in its first year. Similarly, in a downturn, a contractor may appear to have the protection of contract completions to postpone the time when profits fall, the reduced ability to spread overhead may bring this forward.

Claims and variations

Just as there can be confusion between the terms 'claims' and 'variations', so there can be on their accounting treatment. Insofar as a variation is a change within the terms of the contract, accepted by both parties, then prudent credit would normally be taken for the amount of the variation, even though not finally agreed. In contrast, claims are, in the words of SSAP9, 'subject to a high level of uncertainty regarding the outcome' and prudence dictates that credit is not taken for them until they are agreed. Only a minority of the large contractors actually state a specific policy regarding claims in their report and accounts. Of those that do, Taylor Woodrow does not take credit for claims 'until cash is received'; Wimpey and Balfour Beatty will take credit when the claim is, respectively, 'certified' or 'agreed'; while Birse used the expression 'agreed in principle'. There is no suggestion that contractors who do not specifically state their accounting policy on claims are any less conservative, although SSAP9 does give a little more latitude than the 'when agreed' or 'cash received' standpoint:

> . . . it is generally prudent to recognise receipts in respect of . . . claims only when negotiations have reached an advanced stage and there is sufficient evidence of the acceptability of the claim in principle . . . with an indication of the amount involved also being available.

In other words, credit can be taken under the Standard for some element of claims before they are formally agreed. SSAP9 necessitates a subjective judgement as to what is an 'advanced stage' and 'acceptability . . . in principle'.

Retentions

While a contractor may have a claim against a client, the client will also be holding money back from the contractor. This sum, typically up to 5% of the contract price, is know as the 'retention' and is held back for a pre-determined period, in case of latent defects. Although there will be instances where the contractor knows there will be remedial expenditure, this is accounted for in the normal course of the contract profit calculation where the contractor takes into account 'not only the total costs to date and the total estimated further costs to completion...but also the estimated future costs of rectification' (SSAP9, Appendix 1, p. 25). The retention itself will be brought into the profit and loss account for full value and will be held for that amount as a debtor in the balance sheet.

Balance sheet implications

The discussion of work in progress has concentrated on the profit and loss implications, as that is where the greater sensitivities lie, and where the financial analyst will be concentrating his efforts. However, some comments on the balance sheet ramifications may be appropriate. Browse through an old set of contractors' accounts and the largest single figure would often be the gross value of work in progress, perhaps several times the size of annual turnover (Wimpey's 1988 accounts showed contract work in progress of £2127m against contracting turnover of £620m). From that gross figure would be deducted the progress payments received, leaving net work in progress usually less than 10% of the gross (in the Wimpey example, the net work in progress was only 3%).

Following the 1985 Companies Act, which led to a conflict with the requirements of SSAP9 (see Capon, p. 230), the Accounting Standard was revised (September 1988) and the 'amounts recoverable on contracts' is now recorded as a debtor, the argument being that it is an amount which is contractually due. The 'amount recoverable' is defined as the amount by which recorded turnover exceeds progress payments received and a gross work in progress figure is no longer shown.

Where payments on account exceed turnover to date, then the excess is disclosed separately within creditors. These payments will tend to be increased by the prudential approach to profit taking, particularly the requirement to provide for all losses when foreseen, i.e. well before the additional costs are incurred. A delay in recognising profit, by increasing creditors, helps to create the negative capital employed disclosed by some contractors in their segmental analysis of net assets.

The development and interpretation of construction accounting issues and cash flow can be found in the chapter on profits forecasting (Chapter 10).

HOUSEBUILDING

In comparison with contracting, housebuilding accounting might appear relatively straightforward; at its simplest, a single dwelling unit is sold, the money comes in on the day of legal completion, and the profit is taken accordingly. The first qualification to that opening sentence (which is meant to do no more than emphasise the contrast between the three bedroom semi-detached and, say, a hospital) is that not all housebuilders take profit on legal completion of the sale, i.e. the day the building society pays the cheque to the builder and you move in. Some housebuilders take profit on exchange of contract, providing the house is 'substantially complete'; in doing so, they do have accounting logic on their side as they correctly recognise the legal reality of the transaction rather than the final collection of the debt. However, legal completion has become the norm for listed companies, investors preferring its greater conservatism and the lack of subjectivity involved in determining 'substantial completion'.

Having chosen to start this section by stressing the relative simplicity of the product, it still has to be indicated that housebuilding is not without its own accounting complexities. In part, these stem from the fact that houses, although sold on an individual basis, are built in groups on a common site, and their completion may straddle financial years. One could actually argue that this makes the housing site more complicated than the earlier example of a hospital for, although each represents a single construction project, the former could have over a hundred separate purchasers rather than just the one. Thus, the housebuilder has to allocate site costs to individual units, as well as value work in progress at the end of the financial year. In addition to site accounting, the housebuilder has to face a number of issues which are largely peculiar to his industry and to the development world in general: the costing of marketing incentives, land provisions, capitalised interest, and so on.

Allocation of costs

Housebuilding accounts would normally be drawn up on a site by site basis; larger sites would be broken down into phases, perhaps linked to the construction of a road, or to a particular geographic feature of the site, or to a marketing decision about sales rates. When individual units are sold, an allocation of non-build costs has to be made. Not the least important of these is the land cost. On a simple site, where all the plots are the same size, no more than a division sum is required. Even then, some positions could be better than others and housebuilders may load site costs proportionately onto the earlier sales, arguing that the more difficult plots are those which tend to end up being sold last, and these should not be burdened by the same plot cost as allocated to the better sellers.

Just as important are the costs of preparing the site – grading, roads, sewers, services, etc,–and these have to be apportioned to each phase or individual unit. Some housebuilders operate a formula whereby the site costs are loaded more heavily in the earlier phase of the development making, other things being equal, the latter part of site the most profitable. Some costs may not even be known, such as a golf course or school put in at the end of a large development.

More complicated issues arise where a site is used for more than one purpose. Suppose a twenty acre site has five acres sold at a premium price for a supermarket; is the totality of that sale to be deducted from the book cost of the site, leaving the remaining housing land in the books at reduced cost? This could be justified if the housebuilder had planned his purchase on the basis of a potential supermarket sale and incorporated the calculation on added value into his original bid. Or might he argue that the excess profit realised from the sale of land to the supermarket was purely fortuitous and that the full surplus over the average cost per acre should be included in the year of the supermarket sale?

During the course of the development of a long term site, costs and selling prices will inevitably differ from the earlier estimates, and future estimates will also be changed. These changes can have quite different effects on the profits in the year they are made, according to the accounting policy adopted. Assume that on a three year site the appropriate forecasts are made for selling prices, site infrastructure and build costs, and whatever contingencies deemed appropriate are built into the forecast. The estimated profit on the site is £600 000 and in the first year one third of the units are legally completed exactly to budget; a profit of £200 000 is accordingly taken. During the second year another third of the units are sold but costs rise or prices fall to the extent that the estimated profit for the whole site falls to £450 000.

	Profit Allocation (£000)		
Years	*1*	*2*	*3*
Expected profit on whole site	600	450	450
Annual profit taken			
Method A	200	125	125
Method B	200	100	150

Method A would put a line across the profit that had been taken previously; identify a remaining profit of £250 000 over the last two years of the contract and take the due portion, one half, in year two i.e. £125 000. Year three would also see a profit of £125 000 if there were no further changes in the assumptions.

Method B 'cost to complete' would produce a different answer. In year two, it would take the revised estimate for the whole site of £450 000 and argue that two-thirds of this should be taken by the end of year two, i.e. £300 000. However, as £200 000 has already been taken in year one, that leaves only £100 000 to be booked in year two; if there are no further changes in assumptions the profit in year three would rise to £150 000. The use of 'cost to complete' accounting on a long term site accentuates the movements in profits in years when assumptions about future prices and costs are being changed.

What is the house price?

During recessions a whole new range of trading incentives appear, all of which have their own accounting problems, compounded by the treatment of falling land values. The accounting treatment is that these incentives are regarded as cost of securing the sale, normally treated as part of the marketing expenses. This accounting treatment, logical in its own right, does sometimes lead housebuilders to confuse the contract sale price of their product with the effective price. Thus, one may hear statements to the effect 'we do not need house prices to rise to increase our profits, we only need to reduce our marketing expenses'. However, in economic terms the incentives to prospective purchasers to induce them to buy are logically a reduction in the net proceeds received; indeed, there is often an alternative cash discount which would be a deduction from the selling price.

Some of the incentives can be costed accurately at the time of sale and accounted for accordingly, an offer to pay stamp duty or solicitor's fees, for instance. Other incentives may be contingent on a future unknown, a house price or rate of interest being the most common. Some of these contingent costs can be crystallised as a current expense, either through an insurance premium or some form of hedging. Descriptions of the incentives and marketing aids which can cause the most debate in accounting circles follow.

Part exchange

Whereas part exchange is considered the standard method of trading in car showrooms, there remains a body of opinion within the house-building industry which regards part exchanging houses with deep suspicion and, if to be used, only as a last resort. Firms such as Barratt, however, have used it as an integral part of their marketing for twenty years. Housebuilders do not normally include the subsequent sale of part exchanged houses within turnover, preferring the logic of identifying the principal transaction to the accounting practice of recording all sale receipts as revenue.

When taking profit on the sale of the new house, the two extremes would be to argue that until the part exchanged house is sold, the transaction is not completed and no profit on the new house should be taken. Alternatively, the part exchanged house can be assumed to have the value paid and the full profit on the sale of the new house is taken. In practice, a range of treatments within these extremes is adopted, with charges being made for the assumed cost of holding the part exchanged stock, the cost of marketing the part exchange and sometimes a provision against any potential loss on sale. Clearly, when the part exchange house is sold within the same accounting period as the new house, the full outcome is known; in effect, the problem for the statutory accounts becomes one of year end valuation of the part exchange stock plus provision for any future costs.

111

Shared equity

The risks and rewards can vary enormously according to the structure of the shared equity. Typically, the developer will retain a proportion of the equity in the dwelling, up to 50%, with provision for the purchaser to buy the developer's share in future years on an agreed basis. The shared equity may be interest- or rent-free, or a rate of interest may be paid which in itself can be fixed, linked to an index, and/or be below market rates. The rights and obligations to purchase the balance of the equity may be optional or compulsory and the terminal date will vary.

The shared equity will always allow the developer to participate proportionately in any increase in the value of the house and many share also in the downside risk. There are others that effectively are upwards-only equity schemes in that they are structured as a loan agreement with a minimum repayment of the original principal, plus any share in the subsequent appreciation.

The accounting treatments of shared equity range exceptionally widely and have the potential to make a more significant impact on stated profitability than virtually anything in this list. The normal accounting treatment is to credit 100% of the sales proceeds and transfer the shared equity to the balance sheet as a deferred debtor. The accounting decision then becomes how to value that deferred debtor. The ultra-conservative. view, typically found in private companies with more concern over the tax burden than the cosmetics of the profit profile, is to write the shared equity down to a nominal £1; thus, a 25% shared equity on a site which was yielding a margin of 25% would eliminate the whole of the profit. The other extreme is to leave the deferred debtor at its face value, thus taking the same profit on the house at if it had been sold in the conventional manner. This has to be wrong for, if the developer had been able to sell the whole of the house at its face value, then there would have been no need to resort to a shared equity.

Where the interest payment on the outstanding equity is below market rates, a discounted value calculation would immediately require a write-down which, in the case of nil interest equity, would not be insubstantial. There would also be discount for unmarketability.

The value of the full shared equity will rise and fall over time according to the value of the underlying house but, because of the existence of the mortgage as a first charge, the potential for loss can be greater than the potential for gain. Take a 20% shared equity scheme, where the houseowner has a 100% mortgage on his 80% equity. If the house price rises by 25% then the developer participates in a 25% appreciation on his asset. If, however, the house price falls by 25%, the owner of the shared equity now has an asset which is worth, theoretically, 75% of the original cost. However, the total value of the house is now below the amount of the mortgage and if the houseowner has not been able to meet payments, and a repossession ensues, then, as a second charge, the owner of the shared equity receives nothing.

Loans

Some developers advance loans at low or nil rates of interest to prospective purchasers to enable them to cover the shortfall on the deposit requirement. For reasons similar to those covered already it is not prudent to carry these loans at face value and a provision should be made at the time of sale.

Mortgage subsidy

In the USA, with its tradition of fixed interest mortgages, it is common for the developer to purchase a specific interest rate subsidy to cover the first few years of the life of the mortgage; the cost is known and paid at the outset and there are no accounting problems in allocating the cost. With the British variable rate mortgage, the cost of subsidising the first, say, two years of a mortgage at a fixed rate will not be known until the end of the period. Whatever common-sense views may be expressed about future interest rates, there is no theoretical upper limit to the extent of the subsidy and therefore ample scope to argue what might be a reasonable provision. The cost may, however be crystallised at the time of the subsidy by a purchase agreement with the loan provider.

Unemployment insurance

During the recession, unemployment insurance has become a common feature of purchase packages; these are frequently reinsured by the housebuilder at a known cost.

Buy back deals

Although not a widespread practice, there have been instances of housebuilders selling property to Business Expansion Schemes (BES) and underwriting a medium term repurchase commitment.

A sharp increase in the proportion of sales secured by one or more of these sales incentives should be a warning signal for analysts. It could mean that the developer has a marketing advantage over its competition which is enabling it to out-perform. However, it may also be that the developer is securing its increased sales by wrongly pricing its incentives and failing to make full provision for potential future costs in its current profit and loss account.

Housebuilding provisions

Whereas appreciation in land values is credited to the profit and loss account only when the house or the land on which it stands is sold, the Accounting Standard requires that any perceived diminution in value, whether realised or not, must be recognised immediately. Thus, a downward movement in land values is concentrated in a shorter period than is an upward movement.

The Accounting Standard on Stocks and Long Term Contracts (SSAP9) says that 'stocks normally need to be stated at cost or, if lower, at net realisable value'. It is important to stress right at the beginning that what the Standard does not say is 'the lower of cost or *market value*'. If the land holdings were being valued in isolation then 'net realisable value' might be taken to be market value less the associated costs of selling or realising the asset, producing much the same answer. However, housebuilders argue that their land is bought solely for development and the sites should be valued through to completion. Thus, housebuilders do not value their land at its resale value as raw land, nor at what it would cost to replace with equivalent land. Instead, they apply to each *site* the net realisable value, defined by the Standard as 'the estimated proceeds from the sale of items of stock less all further costs to completion and less all costs to be incurred in marketing, selling and distributing'. Thus, if the estimated net proceeds from developing the whole site fail to recoup the cost of the land and build costs (i.e. there is a site loss) then at that stage the land has to be written down. How much does it have to be written down? To the point were there is no longer a loss; in other words, the site will be carried at an estimated zero margin. For the housebuilder, this definition of net realisable value is clearly significantly higher than the market value of the land, as land is only purchased with the intention of securing a given gross return.

The interpretations of net realisable value can vary significantly within the industry. Firstly, to obtain the estimate of total sales value, a judgement of future house prices must be made which can be more open to debate than the costs to complete. There can be differences of opinion as to whether costs to complete should include only site costs or a proportion of administration and overheads. Finally, there are builders who write down the value of their land so as to leave a small positive return rather than the nil return which the Standard might imply.

The argument over the validity of net realisable value or market value is an interesting one and the proponents of market valuation are correct in pointing out that failure to write down to market value can lead to the balance sheet overstating the true current net worth of the business. However, this is probably of more concern to those looking at the break-up value of housebuilders rather than their value as a going concern, but it is important for the analyst to remember that, in extremis, the balance sheet values legitimately produced by net realisable value calculations will not hold good in a forced sale of undeveloped land (or even an orderly one).

The Accounting Standard does, of course, specifically say that writing down stock to estimated replacement cost, i.e. market value, where this is lower than net realisable value, 'is not regarded as acceptable' if the effect is to produce a loss greater than that which is expected to be incurred. However, market valuation of land has been found in the quoted company arena on occasion, where the housebuilder has a stated intention to dispose of some of the undeveloped land rather than to build it out (e.g. Prowting in 1991/92) or on the acquisition of a housebuilding company where the fair value of the victim's land bank is incorporated into the accounts of the successful bidder (e.g. Raine Industries' bid for Walter Lawrence).

The recovery

With this book being written in the wake of unprecedented provisions, concentration on their initial accounting treatment is inevitable. These provisions may turn out to have been accurately assessed but it may also transpire that some companies have over-provided; indeed, if the recovery in housing sales and land prices which has been seen in the first few months of 1993 continues, then the treatment of earlier provisions will become a significant analytical issue. If companies dispose of land and work in progress at prices in excess of written down values, there is no requirement to disclose the extent to which they have clawed back any of those provisions. The quality of the profits recorded in the recovery phase of the cycle will therefore be harder to judge.

Capitalised interest

The practice of capitalising interest, although on the wane, has long been a matter of contention within the property and housebuilding development world. The author does not believe that there is an answer which is right or a wrong in all circumstances; the logic of including the cost of finance in evaluating the profitability of a specific asset purchase and the 'matching requirement' is counterbalanced by the prudential requirement to ensure that consolidated profits are only struck after debiting all recurring charges.

For: the argument for capitalising interest

The argument for capitalising interest (i.e. adding it to the cost of the asset rather than deducting it from profits) is supported by Schedule 4 of the 1985 Companies Act: 'All income and charges relating to the financial year to which the accounts relate shall be taken into account, without regard to the date of receipt or payment.' The commercial logic of capitalisation is best illustrated by the purchase of a single site which is sold many years later without being developed. If the rate of inflation in land prices is exactly the same as the rate of interest over that period, then the reality of the transaction is that no profit has been earned. If interest is not capitalised, but charged each year, then the profits are being depressed by the amount of that interest for a period of years, followed by an artificially high profit appearing to come in the year of the sale. The picture is the same where the land is developed, the developer merely appearing to earn particularly high profits the year in which his old land is utilised.

Most companies, of course, have a mixture of fixed interest finance and equity capital and, whereas internal project accounts can arbitrarily charge all asset expenditure with an assumed rate of interest, for statutory accounts the Companies Act only permits interest to be capitalised where the borrowing is specifically related to the project. It also allows only that interest necessarily incurred during the period in which the asset is being constructed and does not permit the capitalisation of interest charged where the owner has chosen to own the asset for a longer period than necessary. (In the example above of a site held without being developed, the capitalisation of interest would not, therefore, be allowed.)

Against: the argument against capitalising interest

The argument against capitalising interest in statutory accounts is based on prudence, experience, and rests on the implicit assumption that group accounts are drawn up for different objectives than project accounts. Interest is a regular charge which can only be avoided by repaying the loan which, in turn, normally depends upon the sale of the asset against which it is secured. Although interest may be a necessary expense incurred in bringing an asset to maturity, it has no tangible value in its own right (unlike the bricks and mortar). If it is not charged in the accounts as incurred, it is because there is an assumption that the growth in the capital value of the asset will cover the accumulated capitalised interest. To that extent, credit is being taken for future profit before it is secured. There is a particular danger when capitalised interest represents a high proportion of published profits, that dividend payments are made out of profits which have yet to be earned, the company thus becoming a hostage to future movements in asset prices.

In periods of recession, often characterised by sharp downward movement in asset prices, the practice of capitalising interest is normally significantly reduced. In part, this is because the prudential requirement becomes more relevant, but is also because the requirement to value the stock at the lower of cost and net realisable value provides its own natural limit. If realisable values are falling, then there is the possibility that further adding to the cost with another year's capitalised interest might lead to that cost being in excess of net realisable value. At that point, many companies cease to capitalise interest, but there are examples of the logical, though paradoxical, practice of capitalising interest with the left hand and simultaneously providing against the value of that asset with the right hand. A notable defendant of this practice was Trafalgar House.

A fuller discussion on capitalising interest can be found in the Institute of Chartered Accountants' *Accounting and auditing guide: Property company acccounts*, Chapter 4.

Accounting for options

In addition to buying land outright, developers also purchase options on land, acquiring the right to buy land at some future date, usually on the successful granting of planning permission. Typically, a developer will pay a relatively small capital sum for an option on land without planning permission; will use his resources to secure planning permission; and then purchase the land, either at an agreed fixed price or an agreed percentage discount to market value. These options may prove worthless if planning permission is not secured within the agreed time frame; in a falling market, even a successful planning approval may be insufficient to create value in the option if a pre-determined fixed purchase price for the land is in excess of the market price. Conversely, a successful planning application secured in a rising market may generate substantial value.

It is rare to find the accounting treatment on options discussed in published accounts, nor the hidden liabilities which may exist. There could legitimately be a policy of writing off all option payments immediately. Alternatively, options can be held at cost until such time as they can be successfully exercised, in which case the option price would be added to the cost of the land, or until they expired, when options would be written off. As always, perceived diminution of the value of the option should be recognised immediately, although this is a particularly subjective judgement.

The normal understanding of an option is that it gives its owner the right to exercise it if he so wishes. The boom of the late 1980s swung the balance of negotiating strength in favour of the sellers of land. Discounts to market value narrowed, and occasionally even vanished; more serious was the practice – accepted by some housebuilders – of obliging the seller to exercise the option if certain conditions were met, normally the awarding of planning permission, i.e. a put option. In a falling land market, the obligation to purchase at a fixed price, determined in earlier, more optimistic times, could be an onerous liability. Even an obligation to purchase at a discount to market price posed difficulties for those developers at the limit of their borrowing powers. Developers facing an unfavourable put option may prove less diligent in their attempts to secure planning permission, but an option contract which requires them to use their best endeavours could still pose problems.

Conditional contracts have similarities with options in that the substantial payment is still associated with the granting of planning, with full payment being made on the satisfaction of the 'condition', i.e. planning consent. However, unlike the option, title to the land is purchased outright, say at agricultural value, so that if the planning application proves unsuccessful, the developer at least retains an asset.

Land swaps

There is sound commercial logic in housebuilders exchanging land between themselves. A large site will support a higher rate of sales if there is a diversity of product, and one way of achieving this is to bring in more builders. The capital costs of holding long term land is a further incentive for builders to swap 'surplus' land on one large site for land in another area. In those days when market value exceeded the book cost of land, the accounting treatment of land disposals was an important issue, and one which will undoubtedly re-emerge. The contrast between unrealised land appreciation (no accounting profit) and an independent sale of land to a third party (an accounting profit) is clear. Swapping land is a grey area. Normally, the exchange of land of equal value is not treated as a realisation; the tax implications would normally dictate this to be the economic as well as the prudent route. However, if a housebuilder feels the need for additional declared profits then the transaction can be structured so that there are separate sale and purchase agreements. If this was questioned as not being an 'arm's length' transaction, then what about a three-way land exchange? If a housebuilder wants to create 'profit' out of a surplus on his historic land bank, without any diminution in the size of his land holdings, then he will find a means of so doing.

Deferred land payments

Land creditors arise where land is purchased at full price but payment is not immediate. The full price of the land is included in work in progress with the unpaid element, which may represent up to 100% of the purchase price, included in creditors. The deferred payments may be paid to an agreed fixed timetable, or as the land is drawn down for use ('on the drip'). There may or may not be a terminal date by which full payment has to be made, regardless of whether the land has been used.

119

It became fashionable to be concerned about land creditors as the recession deepened for, depending on the terms, they represented a potential cash outflow which might not be recouped in a period of falling sales. Such judgements should only be made in the context of an overall analysis of the balance sheet, for the principle of deferred land payments has to be eminently sensible for the housebuilder. Land is a form of stock which can only be processed over a long period of time; the economics of paying the full price for land well in advance of utilising it can only be justified in inflationary and cash-rich times. The deterioration in housebuilders' balance sheets in the opening years of the 1990s means that the industry will need even greater recourse to payment terms related to the use of the land.

The ultimate in deferred payments is the licence arrangement. The builder never acquires legal title to the land and a payment per plot is made on the completion of each legal sale; this could be an absolute amount or a percentage of the selling price. The land content of the house selling price does not appear in turnover and there are no balance sheet entries in either work in progress or creditors. A licence arrangement may be optional on the builder's part but normally there would be an agreed time period during which the land must be used. In that respect it resembles a lease arrangement but, unlike the lease, the accounting standards impose no requirement to capitalise either the asset or the liability.

*I*ndustry forecasting considerations

This chapter is neither a forecast, nor a definitive guide on how to make one; the first would be out of date by the time this book had gone to press, and the second does not exist, or, at least, not in an accurate form. Rather, it is a guide to those issues which may be taken into consideration when making a forecast, relevant sources both of raw data for the forecasting enthusiast, and other people's forecasts for those who believe that one man's forecast is as good (or as bad) as the next. These are also the reflections of one who earned his living for thirty years by making forecasts – though never with 100% accuracy.

It will soon become clear to the reader that this is not a mathematical chapter. Some readers may wish to apply statistical analysis to the wealth of data available on the housing market and the construction industry; computer packages for curve fitting and regression analysis make this a routine matter for those who have a natural feel for the numerical answers. It is hard to take an intellectual stand against the rigorous testing and quantification of economic relationships which this book does no more than plot on a few graphs, but whether such statistical analysis repays its labour is open to debate. Many of the economic relationships appear to work well for a while but then break down completely at other times. That elusive ingredient 'confidence' (or lack of it) is only observable after the event, yet can distort any of the supposedly obvious relationships, and political direction has its own quite separate input which often defies rational, let alone statistical, analysis. If there is a message to emerge from this chapter, it is that there are no easy, reliable, regular through every cycle, predictive indicators for the construction industry. Hopefully this chapter will at least provide ammunition to challenge those who dogmatically present their forecasts to the second place of decimal, yet also provide some guidance to those who believe that intelligent guesswork, supported by a sound knowledge of the basic sources, will enable them to separate that which is more likely to happen in the future from that which is less likely to happen.

Although new housing accounts for less than 20% of total construction output, more is written on the housing market than the rest of the industry put together. This stems from the weakness in all economic forecasting, of concentrating on those areas for which statistics exist and ignoring those areas which are not easily measured. Moreover, housing affects us all personally and creates a level of interest to which, for instance, warehouse investment can hardly aspire. The housing market also has economic and social implications which extend way beyond the amount of annual new investment.

This chapter is divided into the categories currently used by the Department of the Environment in recording new output and, within each one, the short and long term considerations are looked at together.

HOUSING

The strategic framework

Because of the complexity of the subject, there have already been two chapters on housing: House prices and affordability (Chapter 5), and Long term demand for housing (Chapter 6). The long term demand for housing is an integral part of any long term forecast for the construction industry but it also provides the framework against which the short term fluctuations in the housing market can be evaluated. Thus, a given level of short term housing activity can be considered high or low in relation either to past levels of activity or, alternatively, relative to the concept of a 'base' or 'underlying' level of demand which will be derived from the long term forecast. For example, by the end of the 1960s, annual housing completions were still averaging nearly 400 000, the culmination of the drive to end the post-war housing shortage. Yet the rate of household formation was little more than half that and forecast to fall below 150 000 p.a. The long term household formation estimates were sending a clear and, it transpired, accurate signal to the housebuilding industry. At the time of writing, analysts will be making their judgements on a housing market completing only around 170 000 units compared with an estimated Great Britain household formation of around 200 000. Thus, if the household formation estimates are accepted as a guide, the building industry is not keeping pace with new households nor making any contribution to replacement of the housing stock.

However, before too many conclusions can be drawn from the household formation estimates, a view must also be taken on their reliability. It was shown earlier that although the medium term population forecasts are accurate, the estimates of how that population breaks down into individual households is less so. For instance, the latest estimate of household formation in England for 1996 of 175 000 compares with a 1981 estimate of only 104 000. A common mistake in using the household formation statistics is to assume that they are completely independent of anything else that happens in the economy and can be taken as given. In practice, they can be considerably altered by the level of economic prosperity (or otherwise), especially in the short term. Moreover, once formed, the extent to which those households occupy a separate dwelling, or share, or even have a second home, or have a home of smaller or larger size, again largely relates to the general level of economic prosperity.

Further difficulties with the overall household formation forecasts is that they do not distinguish between owner-occupation or rental, whether public or private. There is a tendency, probably correct, to assume that the public sector housing output is a policy decision, which is not to say that it cannot be related to concepts of demographic or social need (as was the case in the early post-war period). Government policy on public housing is normally set out clearly in ministerial statements (official or leaked), ad hoc White Papers, or in the annual Public Expenditure White Paper (referred to later). As a guide to the long term outcome, the policy statements have been surprisingly helpful, expecially if one remembers that a policy to cut expenditure on housing is usually expressed more elliptically than a policy for growth. From Macmillan's 1951 statement in the House of Commons promising 300 000 homes a year, Labour's 1968 post-devaluation reduction in approvals, the Thatcher Government's rejection of local authority housebuilding, through to the 1989/90 funding programme for housing associations, the Government's intentions have been set out for all to see. Politicians being what they are, increases in spending tend to fall short of the target, while reductions are larger. The forecasting exercise, of course, requires not just the identification of current strategy, but also an assessment of where policy may change. And here again we come back to the macroeconomic assumptions within which any construction forecast is made.

Long term assessments of private housing demand can be made with reference to the total requirement for housing, less the change in the rental stock which may be assumed to be given by government policy. This may be supported by surveys of home-owning intentions, such as those conducted by the British Market Research Bureau. Their 1975 projection of expected tenure in ten years' time proved remarkably accurate (see Table 9.1), but the 1983 projection has proved too optimistic. The implications of the recent phenomenon of negative equity (discussed later), may mean that the latest surveys are even further out.

The implicit assumption of interchangeability between public and private housing, e.g. if the public sector does not provide then the household purchases from the private sector, may hold good for some households, but there is far from complete interchangeability between the different forms of tenure. So far, there has been no mention of private renting where the stock, now standing at 1.7 m units, has been progressively falling since the Second World War. The unchallenged long term assumption has been that, because of the historic burden of anti-landlord

Table 9.1 *Survey of expected tenure in ten years time: Great Britain (%)*

Year of survey	1975	1983	1986	1989
Owner Occupation				
In year of survey*	53	60	63	66
Expected	62	78	80	83
Outcome*	62	67.7#		

* National figures, not necessarily identical with the sample.
 # December 1991.
Source: BMRB reproduced in Building Societies' Association publications.

attitudes, and the tax advantages possessed by owner-occupation, this decline would continue. Again, it must be questioned whether the change in attitudes which negative equity is creating will lead to long term changes in the demand for private rented accommodation.

The difficulties of justifying any particular level of replacement demand were touched on in the chapter on long term housing demand. For much of the 'unacceptable' housing stock, there remains a political or social choice between demolition and renovation, with their entirely different effects on the construction industry. It is unfortunate that the large scale demolition programmes of the 1960s served only to clear the ground for the construction of yet more 'unacceptable' housing. The 1974 Housing Act swung the emphasis towards renovation and preservation of existing communities. Whether this is possible for the post-war high rise council estates is questionable; better choice elsewhere, accompanied by rising standards of living, will lead to an increasing proportion of this element of the housing stock being abandoned.

The short term forecasts for the housing market are judged against the estimates of long term or normal housing demand. Accepting the household formation forecasts (200 000 a year), and assuming the need for an annual replacement of 0.5% of the housing stock (100 000), a base level of housebuilding would be 300 000; vary that according to your own views of social trends and replacement needs.

The private housing cycle

Neither of Britain's main political parties is committed to a substantial expansion of public sector housing. Whatever assumptions are made about the long term housing requirement, it will largely fall to the private sector to supply it. The housing cycle will, therefore, continue to be private driven. To assess the cyclical variations in private housing, one would conventionally look at three main determinants of short term demand, all of which are interrelated: interest rates, house prices, and economic confidence.

Interest rates

Housing is a capital good financed on long term borrowing and therefore peculiarly sensitive to interest rates. However, until the liberalisation of the banking system in the early 1980s and the clearers' attack on the mortgage market, movements in interest rates acted more on the supply of housing finance rather than on the demand. In conditions of rising interest rates, building societies were sensitive to the social and political cost of raising mortgage rates, a sensitivity compounded by their administrative reluctance to make frequent changes. This left them at a disadvantage in competing for deposits and the rate of inflow of funds into the building societies diminished, even becoming an outflow in the first quarter of 1974. This, in turn, led to rationing of mortgages and it was this which effectively constrained potential purchasers' ability to buy houses. Although the more liberalised regime in place since the early 1980s does not preclude lending constraints, interest rates are now more effectively acting on the demand side of the housing equation. They do this by directly pricing the potential housebuyer out of the market, i.e. he cannot or will not pay the required monthly amount, through the lending institutions' own borrowing rules which limit the proportion of income taken by mortgage payments, and through its deterrent value insofar as it signals future changes in interest rates, economic prosperity and house values.

Although the broad way in which interest rate movements affect the housing market can be observed (see Fig. 9.1), the reaction is not always

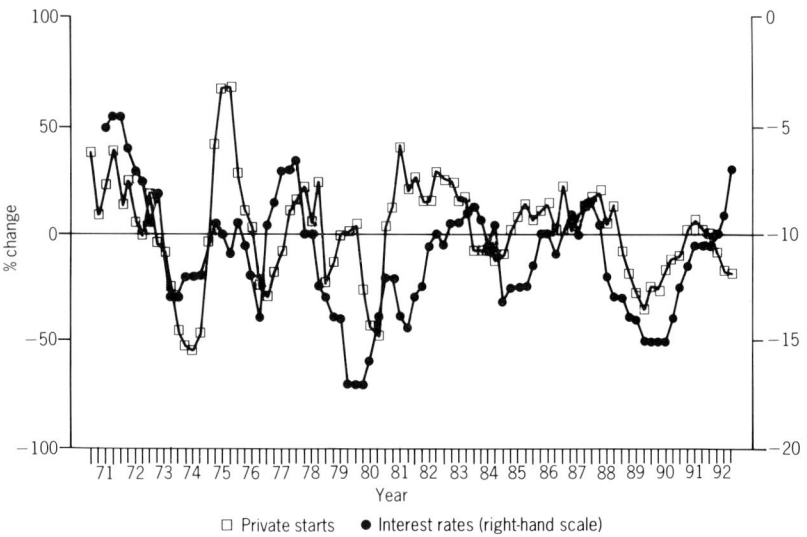

9.1 *Interest rates* vs. *private housing starts.*

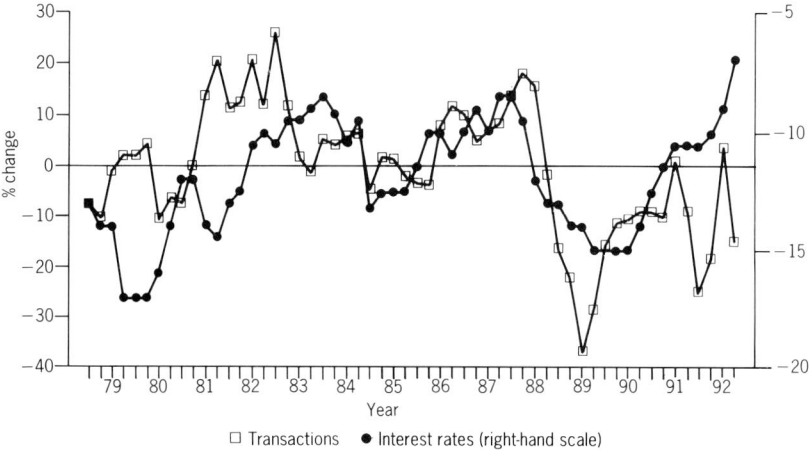

9.2 *Interest rates* vs. *the number of property transactions.*

consistent. An unprecedented period of high interest rates in 1979–80 was accompanied by weak housing markets, especially in 1980, and it can be argued that the partial reduction in base rate during 1980 paved the way for the subsequent recovery in housing demand. However, during 1981, interest rates rose again from their 'low' of 12% to an October peak of 16.5% and this appeared to have no effect at all on what, by then, was becoming a strong housing recovery. Figure 9.2 supports this.

There may also be a distinction between the way in which interest rates affect the housing market on the way up and on the way down. The government has the ability to increase interest rates until such point as buyers are eventually priced out of the housing market. The extent of the rise in interest rates needed to produce a downturn may vary from cycle to cycle but, if they are increased far enough, then the downturn will inevitably result. However, as has been apparent in 1991 and 1992, there is less certainty attached to the government's ability to stimulate the housing market by reducing interest rates. Reductions will mean that a recovery is more likely than it would have been without the reductions, but, if other conditions are adverse, e.g., expectations or economic confidence, then lower interest rates may not be effective. What might have seemed a theoretical point only, proved a practical issue as the housing market contended with the problem of debt deflation at the beginning of the 1990s.

House prices

Chapter 5 set out the sources of the various house price indices and discussed some of the problems of interpretation. One of the first laws of economics that is taught states that, other things being equal, an increase in the price of a commodity tends to reduce demand, and vice versa. In the short term, however, the interplay of expectations can lead to the opposite effect. Insofar as a price increase is an indication of a change in the relationship between supply and demand, it could be the first indication of further changes to come, and the psychology in marketplaces of price increases being used to justify new relationships and, hence, become self-feeding, can be well observed in the Stock Market, classic cars, 'old masters', etc. The house market is no exception (see Fig. 9.3). Admittedly, the house owner does not have the ability to top up his investment in housing in small increments, in the same way that the stock exchange investor can readily increase his investment in quoted equities. However, those who might have been considering entering the housing market for the first time may be stimulated to advance their purchases for fear of having to pay more later; the existing owner-occupier who is considering trading up to a larger property may, similarly, advance his decision. On the selling side, deceased estates may

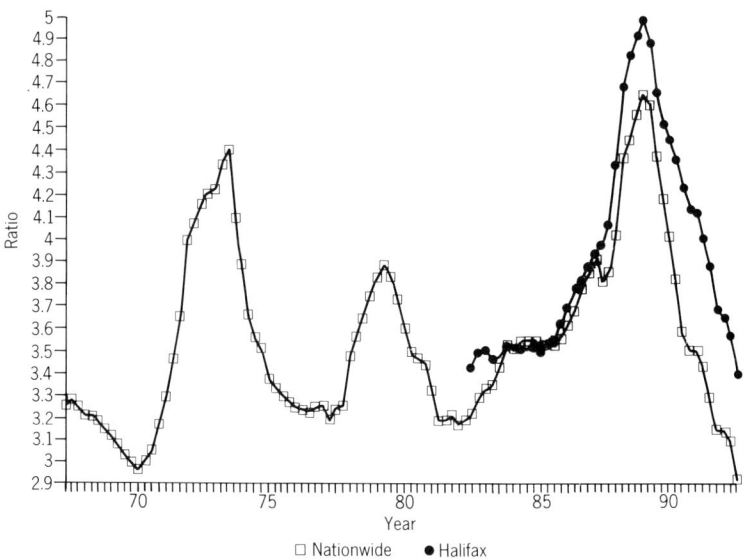

9.3 *House price earnings ratios.*

be more relaxed about delaying sales when time seems to be adding value to their estate. This is no more than a change in the timing of decisions, rather than their quantity, but if this should seem minor in scope, then remember that the total size of the private housing stock is approaching 16 million compared with a private new-build rate of under 150 000; when the annual build is less than 1% of the size of the total stock, then small changes in the preferences of the totality of actual and potential households can have a significant impact on the rate of change in that stock.

Expectations cannot influence prices indefinitely (although at times it may seem like it) and eventually they have to come to terms with the underlying reality of the economics of the market. The upwards limit to house prices tends to be affected by the rules of the lending institution which will typically advance a multiple of the borrower's salary. Those limits may be extended in response to house price increases which have already taken place, and those same institutions are just as much influenced by expectations as their customers. In the end, however, concept of prudence gradually assert themselves and due regard is given to the level of house prices in relation to earnings.

While it is the lending institutions that can act as the most important cap to price rises (if they choose to act responsibly), they have no equivalent power to prevent houses from becoming 'too cheap'. Strictly speaking there is no such thing as a cheap house – they fetch what they fetch. However, there can exist perceptions of what may appear cheap in relation to previous levels of prices, to the cost of building new houses or, more particularly, in relation to current earnings. As most purchasers are fully stretched buying one house, and as houses can only be bought in complete units, the ability of those who believe prices to be cheap to increase their investment by a marginal amount is, as we have observed earlier, not possible. Nevertheless, a response to falling house prices can come in two ways. Firstly, those who could afford to make the purchase at higher prices, but stayed out of the market in the expectation of further falls, may eventually become more concerned about the possibility of missing out on a potential recovery in prices and decide to make their purchase. Secondly, there exists a body of potential buyers whose incomes are such that they were previously unable to afford to purchase, but are able to do so at lower price levels.

Because the inflationary price increases since the Second World War have distorted relative values, the house price earnings ratio plays an important part in the public's perception of whether or not house prices are cheap and there is the natural tendency to use previous historic limits as guides, which can make them self-reinforcing. This book has already argued for caution in using house price earnings ratios for a variety of reasons, not the least of which is that the two main ones do not always produce the same answers. Neither is there any reason why previous peaks should act as limits to either upwards or downwards movements. When house price earnings ratios reached 5.0 (Halifax) in 1988, ample justification was found, including the lower tax regime which, theoretically, enabled purchasers to 'afford' higher prices. At the other end of the scale, in 1992 (Q4) the Nationwide house price earnings ratio breached its previous low point. Will this be rationalised after the event by saying that it is no longer regarded as sensible to over-extend short term finances for the sake of a long term investment gain? Figure 9.4 shows graphically the house price earnings ratios of the Halifax, Nationwide and BSA, from 1983–92.

One of the series which incorporates a number of the factors affecting the affordability of housing is the NHBC 'Ability to buy' index, mentioned in Chapter 4. Figure 9.5 shows the percentage changes in the Inland Revenue transactions series compared with the variations in the Ability to buy index from its long term average of 85. These two lines

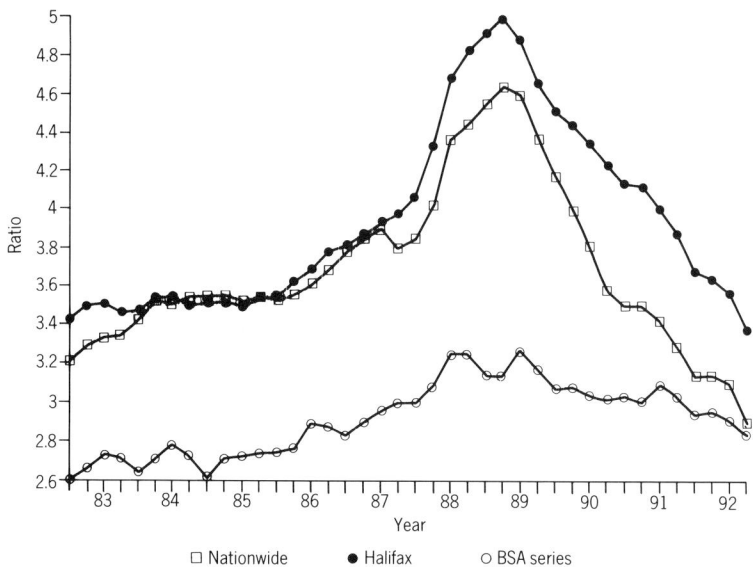

9.4 *House price earnings ratios.*

9.5 *Inland Revenue transactions series* vs. *NHBC's 'Ability to buy index'.*

follow each other reasonably closely, though without the ability to buy index looking like an obvious leading indicator. However, as with so many other indicators, the relationship breaks down in 1991–92, for reasons which are explained in following sections.

Economic confidence

In the short term, purchases of most products have an element of discretion, particularly as to when the timing of that purchase is made. As indicated already, timing of house purchase is affected by the cost of finance and the level of prices – actual or expected. But house purchase is the largest investment made by most households and it requires them to take a view not only on their present ability to finance the transaction, but also requires a view on their ability to sustain those repayments over a long period of time. This relates to their confidence in the economic future of their own family unit and the wider financial environment. There are independent surveys of consumer confidence, but one of the most widely used indicators of confidence in the housing market (apart from housing sales themselves) is the level of unemployment.

131

Table 9.2 *Unemployment compared with housing starts*

				1973		1974
	Q1	*Q2*	*Q3*	*Q4*	*Q1*	*Q2*
Private starts*						
Change, %	+5	+3	−12	−26	−41	−52
Wholly unemployed#						
(000)	555	516	468	421	464	482

* *three quarterly average to smooth fluctuations.*
# *seasonally adjusted.*

The unemployment level is one of those catch all statistics that can readily be used to explain change in demand for any product. When the relationship fits, it seems too obvious to need explanation. In practice, the fit is less than perfect and complicated by the co-relationship between the two variables; housing demand affects unemployment just as unemployment affects housing. Looking at the major cyclical movement in private housing over the last two decades, unemployment was of little predictive value in 1972–75. Unemployment continued falling throughout 1973 (see Table 9.2), well after the housing market had turned down.

The pronounced housing recovery between the fourth quarter of 1974 and the first quarter of 1976 occurred despite a virtual doubling in the level of unemployment (see Table 9.3). Another interesting comparison is from the early 1980s (see Table 9.4). The fall in housing activity in 1980 had coincided with a rise in unemployment from one million to over one and a half million; however, unemployment continued to rise

Table 9.3 *Unemployment compared with housing starts, during recovery*

				1975		1976
	Q1	*Q2*	*Q3*	*Q4*	*Q1*	*Q2*
Private starts*						
Change, %	−3	+35	+59	+55	+36	+15
Wholly unemployed#						
(000)	592	706	820	937	1008	1049

* *three quarterly average to smooth fluctuations.*
# *seasonally adjusted.*

Table 9.4 *Rising unemployment figures compared with housing starts*

	1980				1981			
	Q1	*Q2*	*Q3*	*Q4*	*Q1*	*Q2*	*Q3*	*Q4*
Private starts*								
Change, %	−7	−22	−39	−29	−10	+19	+25	+29
Wholly unemployed#								
(m)	1.05	1.17	1.36	1.62	1.85	2.04	2.18	2.28

	1982				1983			
	Q1	*Q2*	*Q3*	*Q4*	*Q1*	*Q2*	*Q3*	*Q4*
Private starts*								
Change, %	+21	+20	+21	+24	+27	+22	+19	+8
Wholly unemployed#								
(m)	2.35	2.40	2.48	2.57	2.63	2.68	2.70	2.72

* *three quarterly average to smooth fluctuations.*
# *seasonally adjusted.*

through 1982 and 1983 reaching 2.7 m, accompanied by one of the strongest housing markets ever seen.

It can be argued that a high, but stable, level of unemployment tends not to affect the confidence of the 90% or so who remain in employment. This will have been especially true at lower levels of owner occupation when the unemployment was concentrated on those sections of the workforce who typically rented. This was particularly noticeable in the inter-war period when middle class owner occupation came into its own despite appallingly high levels of unemployment. Completions of private unsubsidised housing rose from 66 000 in 1928 to 149 000 in 1932, despite the unemployment rate doubling from 11% to 22%. Private housing completions further doubled to 280 000 by 1935 even though unemployment never fell below 15%

What damages confidence is the thought of unemployment and, therefore, the rate of increase or decrease in unemployment, rather than its absolute level, might provide a better indicator. However, in practice, it seems to offer no great improvement over the gross figures. It can also be argued that the composition of unemployment needs more careful analysis. One reason for the severity of the 1989–92 housing recession in the South East is that unemployment penetrated far deeper into the white collar owner occupiers than it had in any previous cycle.

Debt deflation

Had this book been written in the late 1980s, then the housing section would now be over; there would have been a conventional description of the forces moving the short term housing market since the war: interest rates, house prices, and confidence. The received wisdom for forty years was that house prices will always go up, and therefore, a house should be bought as soon as possible, and should be as large as could be afforded, i.e. borrowed. Two generations of homeowners have derived very substantial equity profits at the expense of the fixed interest depositors who have provided the finance, primarily the elderly and children. In 1973/74 the collapse in the housing market brought with it significant falls in 'real' house prices, but with the general level of inflation well in excess of 20%, nominal house prices changed little. The next comparable fall in the housing market started in the South East in August 1988 and came against a background of low inflation rates; therefore, the burden of falling real asset values was almost entirely translated into falling nominal values. With most first time lending being on mortgages which ranged between 90–100% of original face value, it did not take long for the value of houses purchased recently to fall below the mortgage, creating the phenomenon of negative equity, and an acceleration in the rate of house repossessions.

Calculating the number of households affected by negative equity can be done in a broad brush fashion using the detailed quarterly mortgage and house price statistics that are available. Data exist for the number of mortgages granted, the regional distribution of those mortgages, the proportion of buyers with 100%, 95%, 90%, etc, mortgages and the percentage change in house prices by region. From this, a matrix can be established, showing the percentage surplus or loss on houses purchased in any given quarter after repaying each level of mortgage. Such exercises can be found in the Housebuilders' Federation *Housebuilder*, the *Bank of England Quarterly* and in reviews issued by stockbrokers UBS Phillips & Drew, and (by the author of this book) Credit Lyonnais Laing. The exercises are, of out necessity, simplistic ones, and there are a large number of transactions which hover around the break-even level; choosing one set of house price statistics rather than another may significantly affect transactions at the margin and the Credit Lyonnais Securities' estimates specifically excluded negative equity of up to 5% as being insignificant.

A more interesting question than 'How many homes are affected by negative equity?' is 'What impact will negative equity have on the level of housing demand?' The primary effects of negative equity, i.e. the effects on the specific households concerned, is traumatic, but not necessarily material for the industry as a whole. Those suffering significant negative equity represent a relatively small proportion of the total number of households, and most of those houses do not get repossessed. To the extent that they are forced to stay in their existing house because they cannot afford to trade up, these owners are removed as potential buyers from the housing chain; equally, however, they are also removed as potential sellers to new first time buyers. In other words, they are largely taken out of both sides of the housing demand/supply equation. What will be of much greater concern to the medium term demand for housing is the secondary effect of negative equity on the potential first time buyers who are considering whether to enter the housing ladder. This point was addressed by the author writing in Credit Lyonnais Laing's *Private Housebuilding Annual*, August 1992:

> *The next generation of potential homeowners has seen many of those closest to it in age, perhaps elder siblings, school friends, colleagues who jointed their firm two or three years earlier, facing significant social and financial problems. They can still aspire to home ownership as a long term goal but they may no longer see the need to rush.*
>
> *The householder who decides not to purchase a house for the first time does not automatically become a loss to the building industry if he rents an equivalent dwelling; the financing and tenure may change but the impact on the building industry should be little different; housebuilders are recognising this by their growing commitment to working with housing associations. Once separate households have been formed, typically through partnerships of one form or another, they are unlikely to be dissolved because of changed expectations on house prices; renting a separate home will normally be the alternative to purchase. What is at risk is the rate at which new households are formed. Those most affected here are the young single people whose motivation for buying their own flat was as much the financial motivation of climbing on the housing ladder as soon as possible.*
>
> *It is the category of young single that have the readiest choice of sharing amongst themselves (therefore reduced demand) or living with parents (no additional demand). This grouping currently numbers around 400 000 single male households and around 250 000 single female households in the 15–29*

age range (in practice they must be concentrated in the last ten years of that age band). Those totals were expected to stay broadly unchanged over the next five years, comprising a steady flow of new people entering the category, matched by those leaving it either through marriage or moving into a higher age band. We believe that the housing market is most exposed to the interruption of the potential flow of new single person purchasers. Looking at the broad magnitudes, these must have been averaging 70–80 000 a year. Not all of this will be lost: some will continue to purchase and even sharing will involve the formation of some new households. If the industry loses half the rate of new households formed in the young singles category, the reduction would be well within what the new building market has already forfeited between the 1988 boom and the 1991/92 trough.

The corollary of negative equity as a disincentive to those seeking capital gain, or as a 'frightener' to those who are risk averse, is that the more that house prices fall, the more affordable they become to the aspiring first time buyer. This, in turn, brings us back to the earlier discussions on house price earnings ratios and affordability. These are all factors which must be taken into consideration in forming a view on the likely course of housing activity, but it would be disingenuous to pretend other than that the conventional forecasting relationships were, at least temporarily, thrown into abeyance, leaving a high degree of subjective analysis in their place.

Public housing

The Government's most detailed plans for the public housing sector (and, indeed, its thoughts on private housing) are contained in the DoE's *Annual Report*, part of the annual review of public spending, discussed in more detail under 'public non-housing'. The main capital spending programmes are summarised in Table 9.5.

The Housing Corporation is the public body which regulates the 2300 registered housing associations in England and, as such, central government's capital spending represents grants to the Housing Corporation and not capital expenditure by the housing associations. The Housing Corporation can supplement its resources by borrowing, and the individual housing associations also have access to private finance. Estimates are given by the DoE of the number of housing approvals and completions expected, and these are shown in Table 9.6.

Table 9.5 *Housing: public capital expenditure; current prices (£m)*

	88–89	89–90	90–91	91–92 Outturn	92–93 Est	93–94	94–95	95–96 Plans
Housing Corporation net	738	907	1153	1638	2304	1786	1794	1712
Local authority	1514	1422	1878	2066	1982	1789	1723	1660
Housing Action Trusts				10	27	87	88	90
Total capital	2252	2329	3031	3714	4313	3662	3605	3462

Source: DoE Annual Report 1993, *HMSO, Cmd.2207, Feb. 1993, Fig. 72.*

Table 9.6 *Housing Association development programmes (000)*

	1991–92 Outturn	1992–93 Est	1993–94	1994–95	1995–96 Plans
Approvals					
For rent	46.8	62.5	33.0	36.1	39.1
For sale	4.8	8.1	9.4	13.4	15.6
Total	51.6	70.6	42.4	49.5	54.7
Completions					
For rent	25.5	60.1	44.5	38.0	36.4
For sale	1.3	5.3	10.0	11.4	13.7
Total	26.8	65.4	54.5	49.4	50.1

Source: DoE Annual Report 1993, *HMSO, Cmd.2207, Feb. 1993, Fig. 78.*

The basis on which the DoE determines its capital allocations to local authorities is complicated but was set out in detail in its 1992 Report. The amount that the local authorities can spend on capital then depends upon their own contributions from revenue, their borrowing and the proportion of their capital receipts which the Treasury allows them to reinvest. The reinvestment of capital was constrained by the Treasury during the 1980s but in November 1992 the Chancellor announced that local authorities would be allowed to spend virtually all of their capital receipts accruing between then and December 1993.

The link between central government's capital provision for local authority housing and the eventual expenditure on construction remains tenuous; it can provide a cap to local authority spending, but not a minimum spend. For that reason, the DoE *Annual Report* gives no estimates of the number of houses to be built by the local authorities. In any event, most of the capital spend is devoted not to new building but to significant renovation work (Cmd.2207, Table 5).

PUBLIC NON-HOUSING

Successive privatisations during the 1980s have lessened the relevance of the public sector as an instrument of government financial management (not that that stops politicians trying). For an in-depth analysis of long term construction requirements in the public sector, it is possible to make independent assessments for some parts of the programme. Thus, demand for schools and hospitals can be related to demographic trends or roads to long term transport/miles. Taking schools as an example, one could point to the rise in school rolls during the 1970s, increasing the need for school building, and the corresponding fall in the school population in the 1980s just as clearly indicating the reverse. Unfortunately, many of the long term assumptions remain subjective. Is the standard of existing hospitals acceptable? What is the balance between traffic congestion and conservation requirements? How much should British Rail invest to profit from the Channel Tunnel? One of the greatest mistakes in dealing with long term public expenditure forecasts is to equate what may appear to be perfectly sensible, soundly argued, and socially acceptable estimates of what ought to be done, with a forecast of what the government will choose to do. Overall public expenditure targets are determined within a framework of total expenditure which, in turn, is determined by a combination of the overall rate of economic growth and a policy decision as to how much of that growth (or otherwise) should be diverted to public spending.

The short term framework for government public capital spending is set out each year in a Public Expenditure White Paper (last published in January 1993), and based on the expenditure statement which is made by the Chancellor the previous autumn. During 1993 the traditional Spring Budget for income and the Autumn Statement for expenditure are being synchronised, the Budget of March 1993 being the last at the traditional date. In December 1993 there will be a combined income and expenditure statement, presumably followed early in 1994 by the full expenditure White Paper.

The format of the White Paper is that there is one book covering the totality of the expenditure programme (*Public Expenditure Analysis to 1995–96: Statistical Supplement to the Autumn Statement*, Cmd.2219, Jan. 1993) supported by separate papers for each government department.

The summary tables show the 'estimated outturn' for the financial year nearing completion (1992–93) and the 'plans' for the three succeeding years (see Table 9.7). The data is complex and not always the easiest to use. The term 'capital expenditure' does not always equate to expenditure on assets. It will include capital transfers to the private sector, representing only a proportion of the ultimate private expenditure, and it will be net of asset sales and capital receipts, especially in the housing and environmental budgets. Considerable care must be exer-

Table 9.7 *DoE urban expenditure; current price (£m)*

	87/88	88/89	89/90	90/91	91/92 Outturn	92/93 Est	93/94	94/95	95/96 Plans
City Challenge						64	214	214	214
Urban Programme	246	224	223	226	238	243	176	91	80
City grants	27	28	39	45	41	60	71	71	83
Derelict land	76	68	54	62	77	95	93	93	121
UDCs and DLR	160	255	477	607	602	514	337	293	284
Manchester Olympics Bid(!)					1	13	35	25	–
Inner City Task Forces	5	23	20	21	20	23	18	16	15
Other	–	–	2	3	1	2	7	10	3
Inner cities total	514	598	815	964	980	1014	952	813	800

Source: DoE Annual Report 1993, *HMSO, Cmd.2207, Feb. 1993.*

cised in quoting figures from the central document and, for the largest construction spenders, Environment (including Housing), Health, Education and Transport, the supplementary books should be used if any detailed analysis is being attempted. In particular, analysts should search for expressions like 'expenditure on fixed assets' rather than 'net capital expenditure'.

The DoE *Annual Report* (Cmd.2207, Feb. 1993) is a particularly useful volume for it includes all spending on housing (including new construction, financial support for the Housing Corporation, repair and maintenance) and the plethora of programmes for urban renewal which largely finds its way into public non-housing new work. The urban expenditure programme shows a sharp contrast between the growth which took place up to 1992/93 and the retrenchment which is to follow.

When government policy remains constant which, despite conventional cynicism, can sometimes be for periods of years, the public expenditure capital forecasts tend to be reasonably accurate. If anything, they are subject to slight 'underspend' as it is always possible, for technical or operational reasons, for programmes to fall behind schedule, but the opposite is harder to achieve. If, for instance, there is a programme to build five hospitals, then one can envisage circumstances in which only four are built, but even the public sector will not accidentally build six. Policy changes are, of course, an entirely different matter but these will relate to macroeconomic forecasting assumptions.

Figure 9.6 shows the relationship between the percentage change in

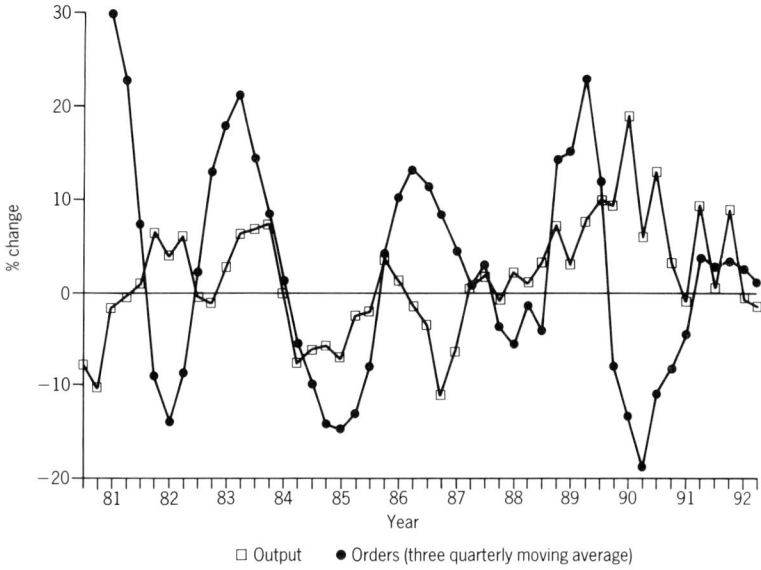

% change

Year

□ Output   ● Orders (three quarterly moving average)

9.6 *Public non-housing output and orders.*

output and orders (the latter smoothed over three quarters). It shows a tendency for the orders to change more significantly than output, except perhaps for the most recent period where there has been a modest increase in public non-housing output, without the benefit of any increase in orders.

PRIVATE COMMERCIAL

Private commercial output (see Fig. 9.7) is affected by many of the economic forces that also determine private housing demand: in the long term, net population growth, internal migration and rising standards; in the short term, interest rates and the general level of economic activity. Interest rates are particularly relevant in that not only do they influence the level of demand for the underlying space, be it office, retail or whatever, interest rates are also crucial to the calculations that drive the speculative development – the financing costs and the yield basis on which rents are capitalised to determine the sale value.

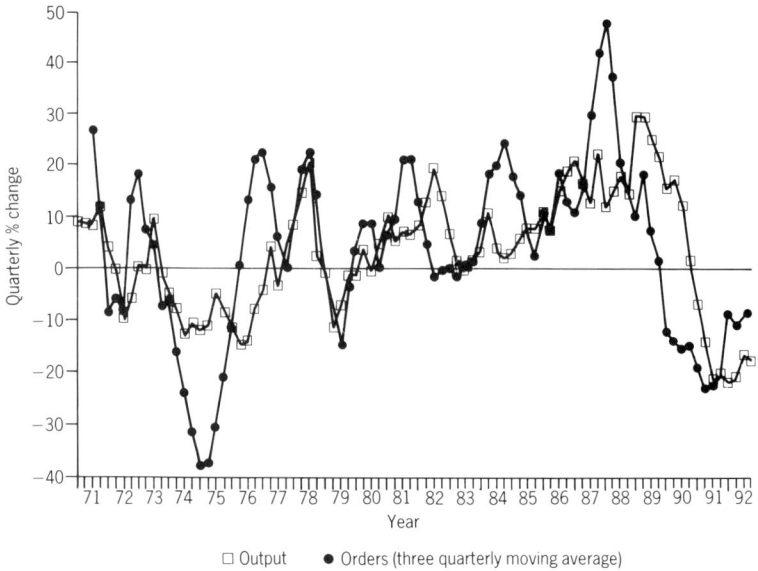

□ Output   ● Orders (three quarterly moving average)

9.7 *Private commercial output and orders.*

The lead times in commercial construction are lengthy and anecdotal evidence will usually be forthcoming as to whether the volume of work in the planning and negotiation stage is either increasing or decreasing. It might be expected that the length of the construction cycle meant that order statistics gave ample warning of changes in commercial output. In fact, the change of direction of output often appears to take place at the same time as orders, which is not a lot of predictive help. Orders are always more erratic than output (even though we have used a three quarterly moving average for the former) and they have from time to time signalled upturns in commercial output which have failed to materialise, e.g. early 1973, 1976, and 1984. The clearest occasion on which orders have unquestionably served their purpose as a lead indicator has been the most recent collapse, when the three quarterly order intake turned negative in the first quarter of 1990 (the first quarterly drop was actually in the third quarter or 1989), but it was not until the first quarter of 1991 that there followed a year on year fall in output.

The property boom of the late 1980s was characterised by very large projects in London, including Canary Wharf, and associated with the wider access to the securities markets. At their peak in 1988, Greater London commercial construction orders accounted for 38% of the GB total, against only 27% five years previous, and 25% in the 1973 order boom. The late 1980s order inflow will also have differed from earlier commercial work in the high proportion of finishing trades, and not only the more sophisticated air conditioning systems, but the communications networks installed in the modern dealing rooms. The lag between orders and output which appeared in this last cycle may, therefore, be less pronounced in future cycles.

The surveys produced by commercial property estate agents were mentioned in Chapter 4, the chapter on construction statistics. They are generally designed to give a view at a given point in time rather than provide a long series, but Table 9.8 combines the Jones Lang Wootton statistics and the DoE construction output for the commercial sector.

The interesting feature of Table 9.8 is that take-up of London office space began to fall sharply at the beginning of 1988, a full year before the first fall in commercial construction orders in the Greater London region.

PRIVATE INDUSTRIAL

Until the mid-1980s, industrial construction output maintained a close relationship not only with industrial orders but also with, as would be expected, other indicators of manufacturing investment such as capacity working and the CBI's *Survey of Investment Intentions*. If anything, and this must be considered an improvement on other sectors of the construction industry, one can see slightly more evidence of the order statistics acting as a lead indicator for the output, just squeezing ahead, for instance, in 1976 and, again, ahead of the major upturn in 1983/1984. The statistics produced by King, Sturge & Co. on industrial floor space availability might be regarded as a determinant of speculative industrial building, but probably have little influence on the substantial industrial building for companies' own use; floor space figures are more erratic than many series and the change of direction appears to occur at much the same time as for output itself. Figure 9.8 shows the relationship between industrial output and orders, but has been stopped at 1986 for the reasons that follow.

Despite a decline in all the leading indicators since the late 1980s,

Table 9.8 *Central London office space (sq. ft., 000)*

		Take-up of space %	Change, %	Development starts %	Change, %	Availability	Change, %	Commercial* construction orders, Greater London Change, %
1983	1st half							
	2nd half	3 900						
	Year			3 680		8 040		
1984	1st half	4 200						26
	2nd half	3 950	1					41
	Year	8 150		4 160	13	8 310	−6	33
1985	1st half	5 920	41					36
	2nd half	5 910	50					10
	Year	11 830	45	4 520	9	7 550	−9	22
1986	1st half	6 570	11					11
	2nd half	5 725	−3					45
	Year	12 295	4	4 880	8	5 050	−33	27
1987	1st half	6 450	−2					38
	2nd half	6 300	10					90
	Year	12 750	4	13 790	183	6 140	22	66
1988	1st half	5 010	−22					87
	2nd half	4 070	−35					4
	Year	9 080	−29	10 700	−22	7 570	23	35

Table 9.8 Continued

		Take-up of space %	Change, %	Development starts %	Change, %	Availability	Change, %	Commercial* construction orders, Greater London Change, %
1989	1st half	4 440	−11					−11
	2nd half	4 580	13					−37
	Year	9 020	−1	8 510	−20	13 500	78	−23
1990	1st half	4 100	−8	4 100				−15
	2nd half	3 790	−17	2 865				6
	Year	7 890	−13	6 965	−18	21 640	60	−7
1991	1st half	2 400	−41	1 330	−68			−51
	2nd half	2 725	−28	530	−82			−51
	Year	5 125	−35	2 160	−73	25 780	19	−51
1992	1st half	2 860	19	330	−75			−30
	2nd half	3 225	18	180	−66			−22
	Year	6 115	19	510	−73	24 385	−5	−25

* Current prices; 1992 figures derived from new orders series.

Source: Jones Lang Wootton/DoE.

145

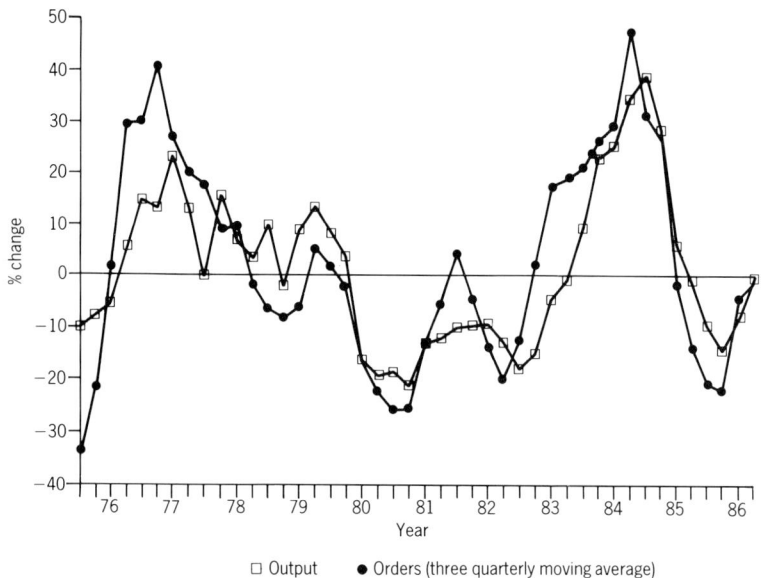

□ Output   ● Orders (three quarterly moving average)

9.8 *Industrial output and orders.*

there was little apparent change in the subsequent course of industrial construction output. This is because there has been a material change in the composition of the industrial building sector. There was, of course, the award of the initial Channel Tunnel contract in the third quarter of 1987, which bore as much relationship to the original concept of industrial building as does a garden shed. On a more permanent basis, the privatisation of what were previously public sector utilities, often with investment criteria determined in a longer term time frame, has broken the old relationship between manufacturing activity and industrial investment; for instance, in the first quarter of 1992, water and sewerage orders were in excess of those for factories. If the output figures relating only to factories and warehouses are extracted from the 'industrial' construction sector, then it can be seen that the traditional relationships have broadly continued for this element of the sector, despite the substantial inflow of overseas manufacturing investment from Japan and the USA. When revised DoE analysis of industrial construction is available for both orders and output, the conventional industrial building cycle will be easier to follow.

The CBI produces useful surveys of investment intentions and capacity working (*Industrial Trends Survey Quarterly Full Results*, Confederation of British Industry) and their survey results can be plotted against DoE factory and warehouse construction output. The balance between those firms working above and those working below capacity does not relate to the short term fluctuations in industrial building, but one could argue that the upturn in net capacity working, which began in 1981 and continued through to 1988, was the basis for the substantial increase in industrial construction work between 1983 and 1988. The fall in net capacity working from 1988 onwards also appears to have some predictive value.

The other CBI statistic which is interesting to look at is the *Survey of Investment Intentions* which is available for total investment and specifically for buildings (see Fig. 9.9). Again, taking the net balance between those intending to spend more and those intending to spend less, the relationship between the net investment intentions and the actual construction output figures can be seen. One of the interesting features of the net investment statistic is that since the series was started, only briefly has there been a time when industry actually expected to invest more rather than less. But it is not the absolute size of the figure which matters, but the change over time and, again, the rise in net investment intentions from 1980 onwards does appear to have signalled the subsequent recovery in industrial construction.

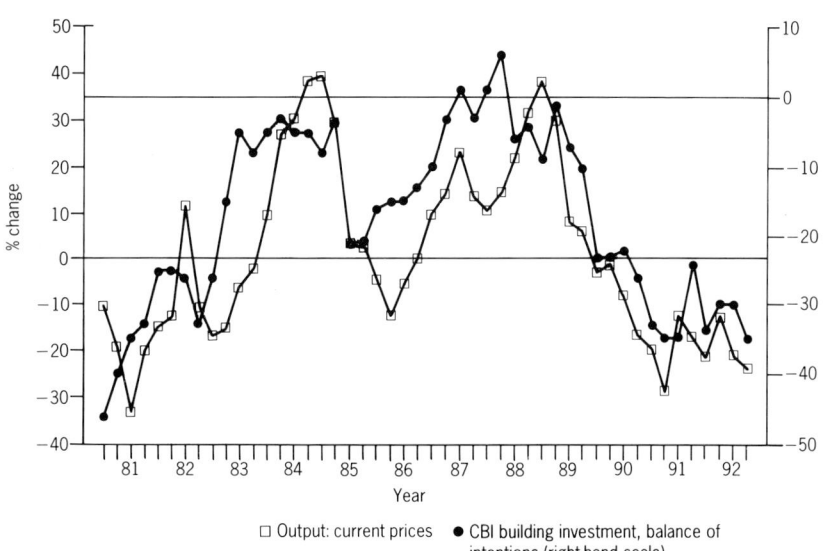

9.9 *Factory building* vs. *CBI building investment data.*

Table 9.9 *Infrastructure orders; current prices (£m)*

	1989	1990	1991	1992
Water	251	321	509	
Sewerage	332	491	429	
Electricity	271	187	301	
Roads	1226	1425	1415	
Other	880	767	769	
Total of which	2960	3190	3423	
Public	2369	2029	2090	
Private	592	1161	1133	

INFRASTRUCTURE

As should now be clear, the DoE has recognised the distortions created by privatisation and is in the process of introducing an 'Infrastructure' category. The order statistics are now available and a breakdown by category is available in current price terms (see Table 9.9).

The largest single component of the new infrastructure category is roads, and the comments made in the Public Non-Housing section (see page 138) are probably still relevant to this politically sensitive area. Not far behind, and by far the largest part of private infrastructure spending, is the capital expenditure programme of the privatised water companies. General background to the long term capital programmes can be found in the public expenditure white papers before privatisation; the original privatisation prospectus of November 1989; and subsequent reports from OFWAT, the industry's regulatory body. The OFWAT reports include one specifically on capital investment and finance (*1991–92 Report on Capital Investment and Financial Performance of the Water Companies in England and Wales*, HMSO) which featured an investment projection up to 1994–95.

REPAIR AND MAINTENANCE

Housing

One of the recurring myths in the construction sector, and recurring in the sense that it is trotted out at the early stage of each recession, is that 'repair and maintenance is recession-proof'. The logic, at first sight, seems reasonable as it is assumed that the day to day essential work of maintaining the fabric of one's capital assets continues and, indeed, the author can swear to having heard perfectly sensible people arguing that rising unemployment is good for the repair and maintenance sector as the unemployed can no longer afford holidays and instead use their time to work on their house. There are two serious misconceptions with this view of repair and maintenance (apart from the obvious idiocy) which must be addressed. The first, of course, lies in the very title 'Repair and maintenance'. As was made clear in an earlier chapter, a substantial element of the work (particularly in housing) is major improvement work, in many ways akin to capital expenditure by corporate bodies or large-ticket consumer expenditure on the part of the individual. If the repair and maintenance sector was, instead, headed 'Medium-sized capital expenditure and large-ticket consumer spending', then the whole analysis of this, the largest component of construction expenditure, would start psychologically from a different viewpoint.

The fallacy of non-postponable maintenance

The comments above are not to deny that there is still a significant maintenance content to the sector, but here we come to the second error, or the fallacy of non-postponable maintenance. While it is accepted that new building can be postponed, for all the various reasons that determine the cycle of new building work, it is argued that 'maintenance', because it is physically necessary, will actually be carried out. Most maintenance is, however, if not infinitely postponable, at least postponable for a sufficient number of years to create its own cycle. Should that be doubted, then one must find an alternative explanation for how so much of the nation's housing stock comes to be in a state of either minor or major disrepair. A house that needed repainting after five years merely needs rather more thorough repainting after seven years. The centre of the window frame that needed replacing becomes the whole of the window frame. The roof that is leaking a little becomes the roof that is leaking rather more. And so one can go on. There are, of course, degrees of postponability and one can think of some repairs, for instance broken windows and the central heating boiler in January, which are carried out promptly, in contrast, perhaps, to the nation's front drives which have an almost indefinite propensity to deteriorate without concern to their owners.

149

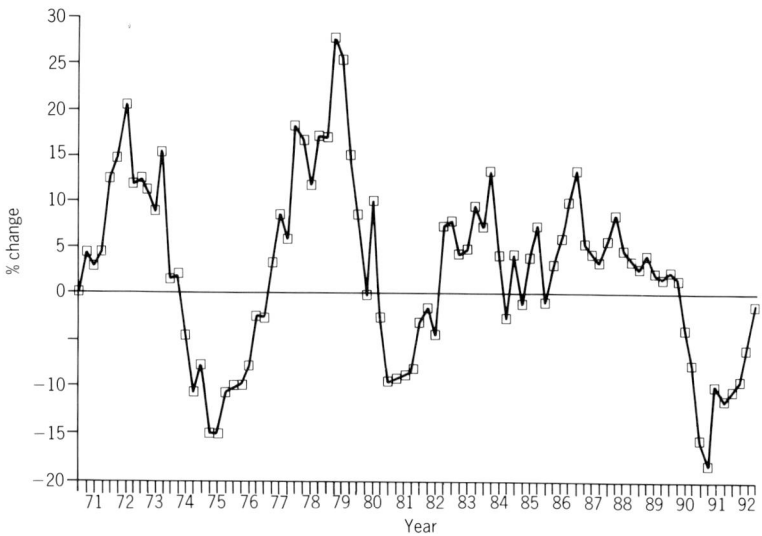

9.10 *Housing repair and maintenance.*

Once it is accepted that the repair and maintenance sector contains elements of capital expenditure (often requiring external financing) and that most maintenance is postponable, then one can look at the repair and maintenance statistics for what they are – cyclical, like the rest of the industry. Figure 9.10 is for housing repair and maintenance from 1970; year on year percentage changes of 10–20% are common, with the year 1979 showing a peak 21% gain and 1991 a 14% fall. To emphasise the point even more strongly, the late 1970s boom in repair and maintenance saw growth of over 60% in three and a half years. The split between public and private housing has only recently been made available with back data to 1985. This shows the two sectors diverging at times, including the last recession, where a recovery in the public sector has taken place before private expenditure.

Although there is only a limited run of private housing repair and maintenance statistics, they do at least show the pattern of one severe recession (see Fig. 9.11). It is logically argued that much repair and improvement work is consequent on house moves but Fig. 9.12 shows that the sharp fall in transactions in 1989 was not immediately followed by repair and maintenance; indeed, it was not until late 1990 that any

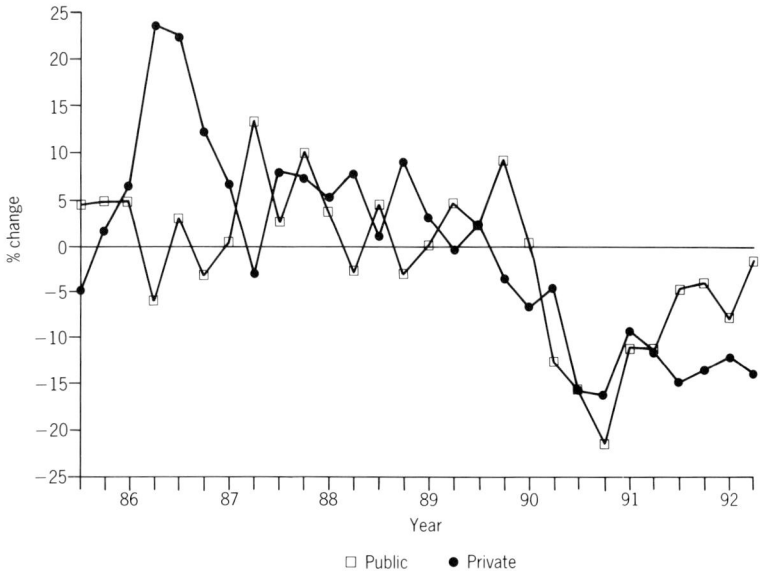

9.11 *Private and public housing repair and maintenance.*

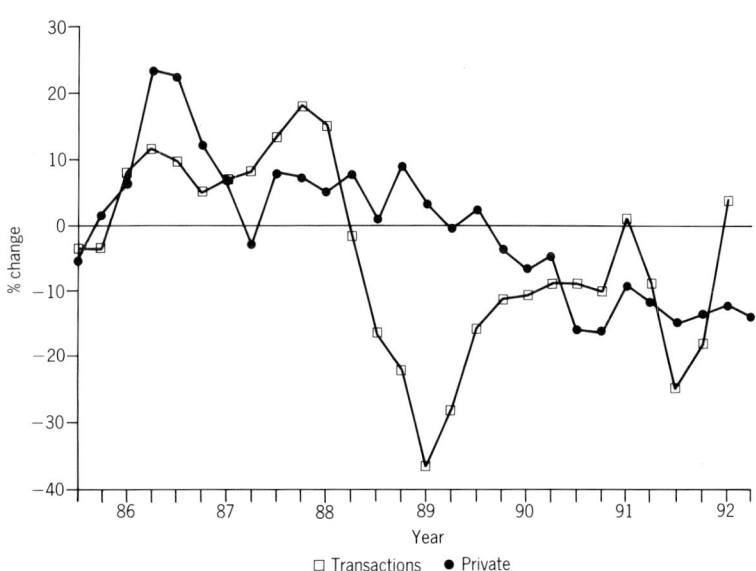

9.12 *Private housing repair and maintenance* vs. *Inland Revenue transactions.*

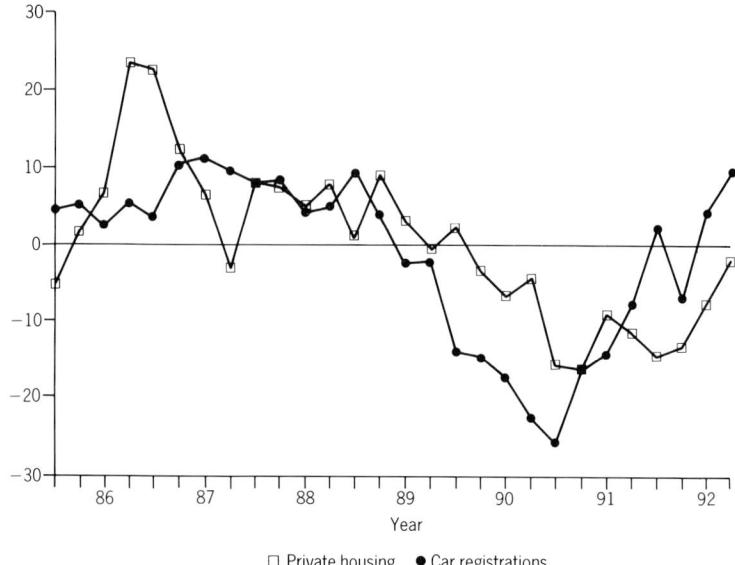

9.13 *Private housing repair and maintenance* vs. *new car registrations.*

significant deterioration took place in the construction data. Looking around at other related statistics, repair and maintenance certainly fell more quickly than consumers' expenditure in total but it is interesting to see that it exhibits much the same profile as car registrations (see Fig. 9.13), confirming the earlier comments about the similarity of the sector with big ticket consumer spending. It should be remembered that the smaller d.i.y. tasks which are inherently less cyclical, will be excluded from the statistical coverage.

As with other areas of the construction industry, public sector policy can produce longer term trends. Thus, the whole thrust of public housing policy since the mid-1970s has been away from new building and towards improvement of the public housing stock. There have also been variations in the incentives given to the private sector, in the form of grants to encourage owner occupiers; this most notably created booms in re-roofing (1982–84) and insulation.

Long term considerations

The long term determinants of repair and maintenance are inherently different from those of new building; the latter is a function of the desired

152

change in the total building stock, whereas repair and maintenance is a function of the overall size of that stock. A long term position can exist where the rate of new building is declining but, as the total stock is still growing, then so will the requirement for repair and maintenance continue to grow. The relationships are similar to those found in the motor industry where replacement demand depends upon the size of the 'car pool' but new car production on the rate of increase in the population of cars. It is only within the housing sector that there are accurate statistics about the stock of buildings but this does cover over half the published repair and maintenance output.

One tends to think of the housing stock changing very slowly over time but, in fact, it has increased by 30% over the last 25 years. However, as the private rented stock has consistently fallen during this period, and the public stock has declined during the 1980s following the 'right to buy' legislation, so the owner-occupied stock has risen – almost doubling over that period. The published statistics do not support this next assertion as all the unrecorded output is concentrated within the private sector, but it is probable that the rate of spend per house is greater within the owner occupied stock rather than the rental stock, reinforcing the growth in housing repair and maintenance.

Non-housing repair and maintenance

Private non-housing repair and maintenance covers the whole industrial and commercial sector. It, too, is a blend of what must be done and what can be afforded, and the accompanying Fig. 9.14 and 9.15 show that expenditure mirrors the basic manufacturing investment cycle. Thus, there was a close relationship with new industrial building until 1991–92 when the privatisation of the water industry altered the mix of the industrial building sector; equally, the graph of changes in non-housing repair and maintenance mirrors the level of capacity working. It could also be argued that the pressure on private non-housing repair and maintenance is related to the cycle of corporate profits and cash flow.

Within the public non-housing sector, the relationship between new building and repair and maintenance is less clear cut. Figure 9.16 shows repair and maintenance at only half the size of new construction at the start of the 1970s, increasing in relative importance during the decade until spending actually exceeded that on new building in 1980, a

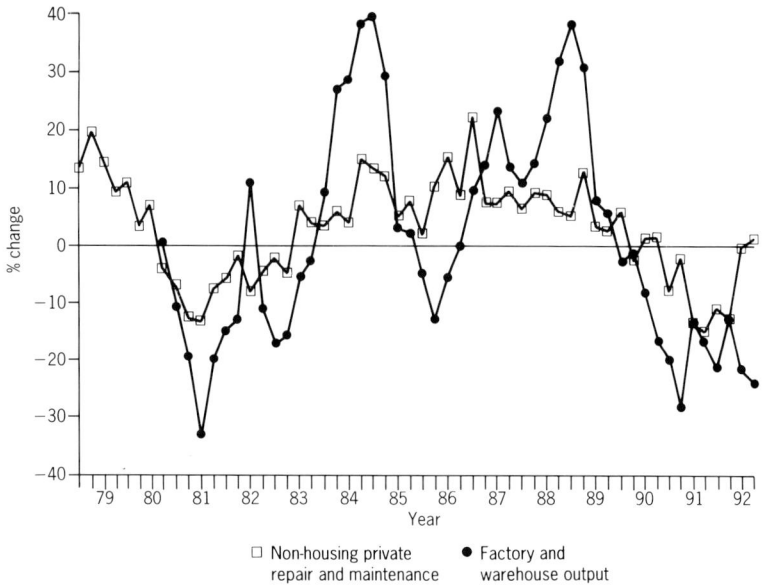

9.14 *Private non-housing repair and maintenance,* vs. *factory and warehouse output.*

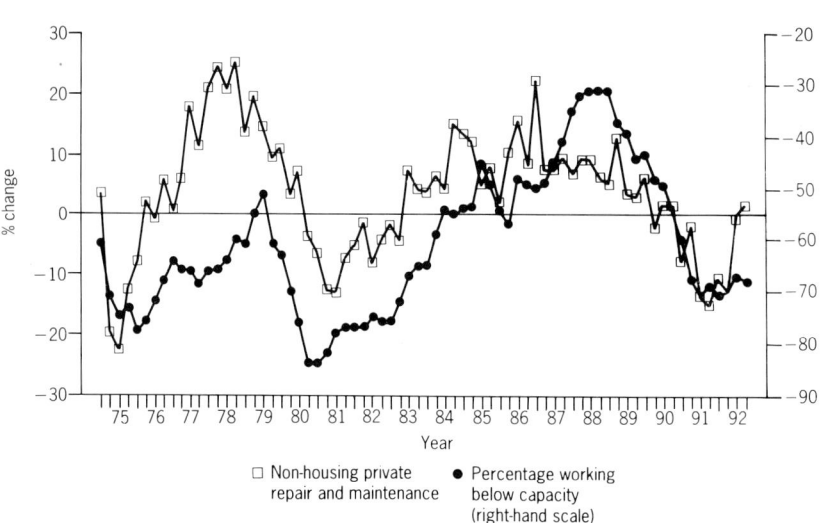

9.15 *Private non-housing repair and maintenance* vs. *level of industrial capacity working.*

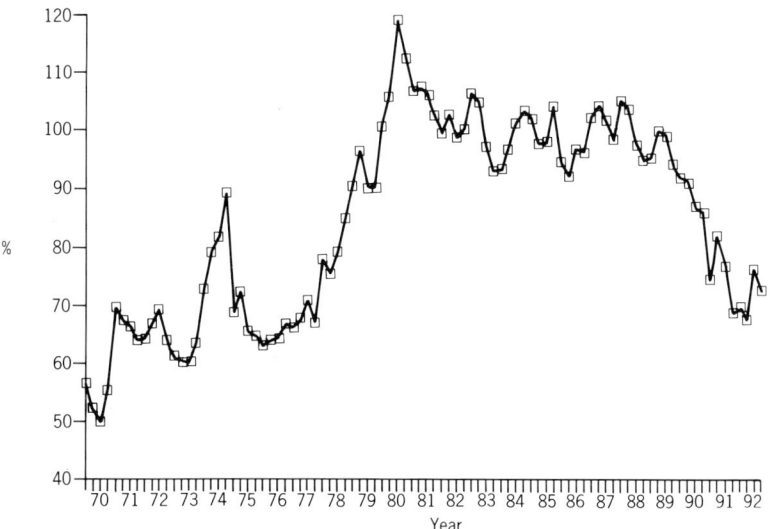

9.16 *Public non-housing repair and maintenance, as a percentage of new construction output.*

reflection of the stringent measures taken to restrict the public sector in the 1970s. More recently, there has been a substantial decline in the relative importance of public sector repair and maintenance. This may reflect the new works being pushed through at a central level, whereas further down the spending tiers (e.g. in the local authorities) day to day financial pressures have placed greater constraints on repair and maintenance. There will also be a statistical side effect from privatisations as output remains within the public sector on those contracts placed before privatisation, whereas repair and maintenance expenditure transferred immediately.

INDEPENDENT FORECASTS

Numerous independent forecasts of construction industry output exist, particularly within the City. Two recognised and regular forecasts by bodies associated with the industry were produced by the Building Materials Producers and the building NEDO (National Economic Development Office); the latter has been disbanded but the forecasts have been continued by Construction Forecasting & Research (Princes House, 39 Kingsway, Holborn, London WC2B 6TP). In January 1993, the widely read *Building* magazine introduced a new two year forecast, compiled by Barclays Bank.

*P*rofits forecasting and cash flow

PROFITS FORECASTING

Introduction

Many of those employed in construction analysis will need to go no further than the industry trends; others, especially those involved in the financial markets, will additionally be concerned with the profitability and share price performance of the constituent companies. References to profit forecasting can be found within accounting and investment literature though rarely touching on the practical aspects – perhaps because profits forecasting is such an inexact 'science' with its mix of industry trends, financial analysis, company briefings, trade gossip and, not least, intuition. The only comments appropriate in this book are to indicate where a construction company might differ from companies in other industries.

The prerequisite of successful profits forecasting is to get the industry forecast right and as we are dealing with a cyclical industry there is a particular emphasis on the scale of the peaks and troughs and on calling the turning points in the cycle. A large part of this book has been concerned with defining the structure of the construction industry; the sources and relevance of the statistical data; and the varying relationships between the component parts of the construction industry and what are regarded (not always correctly) as leading indicators.

It is an unusual industry in which company profits do not respond, to a greater or lesser magnitude, to a sustained rise or fall in demand. In a manufacturing organisation, rising demand leads to greater capacity utilisation and hence lower unit costs; at the same time, that rising demand will also lead to price increases (whether direct or through a reduction in previous discounts). The magnitude of the profit increase created by the double effect of capacity utilisation and pricing will depend on the ability and willingness of the companies to exploit their advantages, and their perceived balance between short term gains and long term strategy. It may be argued that a monopolistic structure maximises the profit gain but, on the other hand, the fewer companies involved, the more likely they are to be able to exercise a responsible long term pricing policy.

Depending on the capacity structure of a particular industry, the duration of the upturn, and the investment policy, companies will eventually run into supply constraints of one form or another which push their costs up faster than selling prices; a reflection of suppliers – be they of labour or materials – seeking their share of the excess profit. Seeing the shortage of capacity and the high returns, some producers will invest in new capacity with an unerring instinct for doing so as their industry reaches the top of its cycle.

The downwards phase of the profits cycle is more or less a mirror image of the above with falling capacity utilisation and price competition. Eventually capacity is closed either voluntarily or through financial pressure. As capacity tends to be removed towards the end of the cyclical downturn when, by definition, the decline in demand is slowing, capacity utilisation can actually increase before demand has begun to rise. There can, therefore, be the apparent paradox of profits improving ahead of the increase in demand.

The construction cycle

Those generalised, and presumably familiar, comments are provided as the framework against which can be set the peculiarities of the construction industry. The first point to be made is that the difference between the profits that are actually being earned, and those that are published can be far wider for construction companies than for any manufacturing and distribution business. This stems from the long term nature of the individual transactions and the consequent inexactitudes and subjectivity in determining the value of work in progress carried forward from one trading period to another. References in the text to profits can be assumed to relate to the reality of the occasion rather than their publicly disclosed version; contemplation of the art of disclosure will be left until later.

The construction industry has its own cycle of capacity utilisation and pricing as described above. However, the extensive reliance on bought in labour and materials, especially the use of the sub-contract system, combined with the long term price commitments on some parts of the workload, means that the profit to turnover relationship is far from simple. The comment was made earlier that the manufacturing company might experience some erosion of profitability at the peak of the cycle due to overheating. This phenomenon can be experienced at a much earlier stage by contractors, especially when they are operating fixed price contracts. Indeed, a contractor carrying out non-housing building work may find that demand in his sector has shown no recovery, yet his subcontract costs are being forced up by pressure of alternative demand from a rising housing market.

The judgement on cyclical movements in costs and prices is one aspect of contract risk; there is additionally the risk attached to the individual contract, irrespective of the stage in the cycle. While it is not unknown for manufacturing companies to face severe problems on one large contract (defence and aerospace manufacturers spring to mind) the frequency with which large contractors can, for want of a tighter definition, be financially embarrassed is peculiar to the construction industry. The next section therefore looks at the wider aspects of contract risk and draws on material contained in earlier chapters on contract structure and accounting.

The description of the different types of contractual arrangement above contains comment on the type of risk and how it may, or may not, be minimised. Because of the uncertainties inherent in competitive tendering, often with fixed prices, and its low profit margins – in reasonable trading conditions a builder might aspire to 2–2.5% margin and a civil engineer as much as 4% – contracting is automatically perceived as being high risk. This does not necessarily follow. Many of the costs and risks of the contract are passed on immediately, either in the form of material purchases or sub-contracts for major parts of work. Thus, the margin on the main contractor's added value, or his own work, will be much higher than it appears from the published accounts. This can reach its extreme in management contracting where the nominal margin in relation to the value of the total contract is at its lowest but the management contractor may be acting only as the agent of the client and, indeed, may not even be writing his own cheques out for the purchases. In that case, his turnover becomes the management fee rather than the total value of the contract.

The area of risk will change materially according to the nature of the market and the position in the economic cycle. In a buoyant market the risk is more likely to be related to cost inflation and the inability to secure supplies or services at the rates budgeted. Even inflation clauses would be no protection to a contractor paying unofficial bonuses to secure labour. Contractors may take a deliberate view of anticipated changes in the rate of inflation in designing the element of fixed price work that they will accept.

In a recession, inflation ceases to be a problem and, indeed, inflation-adjusted contracts will work in the contractor's favour as he secures discounts and rebates which are not incorporated into the official indexes. Instead, the risks become those associated with reducing tender margins to secure new work and the financial security of the client. A mild recession may even favour the main contractor who is able to exercise more downward pressure on the margins and prices of his suppliers as the contract progresses, than was forced on him when the contract was first secured. However, in a deepening recession, this 'benefit' will be more than offset by the fall in the absolute size of turnover and by business failures both amongst the clients and the sub-contractors.

Squeezing the sub-contractor to the point where he fails half way through the contract is of little advantage, but even more damage is done when the main client fails. Insofar as the monthly payments are up to date, the loss should be minimised but failures do not tend to come from amongst the prompt payers and, to the extent that there are perhaps significant claims and variations yet to be paid, the loss can be substantial in relation to the total size of the contract. In the event of a failure of the client, the financial relationship between the main contractor and the sub-contractor can prove a further area of risk and it is currently the source of extensive legal debate. Some contracts between the main contractor and the sub-contractor have what is known as 'pay when paid' clauses; in other words, if the main contractor does not get paid, then the sub-contractor is not entitled to be paid by the main contractor. In other contracts, the sub-contractor is clearly working for the main contractor and is entitled to be paid by him, irrespective of whether the main contractor himself is paid. The legal arguments centre around the nature of these 'pay when paid' clauses, and the extent to which they may be taken as implicit in normal contractual relationships.

If inflation is the risk in a rising market and client failure and falling tender prices the risks of recession, then the contract execution risk remains ever present. The extent to which a contractor can protect himself against the unexpected construction conditions, or the client seeking substantial changes, will depend upon the expertise of the contractor, the structure of the contract, and the nature of the client. Indeed, there are contractors who argue that their most profitable contracts are those which deviate most from the client's original in-structions and assurances. This is true, but that same description could also be applied to some of the biggest losers as well. The ways in which the industry seeks to protect itself have been discussed earlier, but none of the risks are easily analysed from the outside, or at least, not until matters have begun to go seriously wrong.

Balance sheet clues

The balance sheet can indicate the presence of contracts where, for whatever reason, payments are running significantly behind the cost of work executed. The obvious general indicator is a deterioration in the cash position or a rise in borrowings, unexplained by other group expenditure. This will be complemented by a rise in net contract work in progress in absolute terms and as a percentage of contracting turnover. Increases in published profits, accompanied by a deteriorating balance sheet must always serve as a warning shot across the analytical bows. It would be satisfying to think that desk-based analysis could provide all the answers, but sources within the industry will often be more aware of which firms have been taking on contracts at the wrong price, or have specific contract problems. It may of course be professional jealousy but, where one contractor has secured a substantial increase in his workload, and there are both 'noises off stage' and balance sheet strains, then the outsider should have due caution. In a recession, it becomes even harder to evaluate strategy when there are added fears that work with little or no profit margins is being 'bought' for the sake of advanced payments.

A financial model?

As always in this book, the emphasis is on what is specific or peculiar to the construction industry. Nevertheless, there is one general aspect of profits forecasting which appears to attract little attention, and that is the status of the forecasters: are they insiders or outsiders? The 'insiders'– be they directors, employees, auditors or reporting accountants – will (theoretically) have full access to all the corporate information, particularly the cost and pricing structure; model building for them is no problem. For the 'outsider', and this primarily covers investment institutions, little of the detail is available. Some companies have a simple structure, perhaps making one main product on which there is a published selling price and a simulated profits model working through volumes, selling prices and costs can be attempted.

A simplified profits model might be prepared for the housebuilding industry. There is a monthly index of selling prices, land prices are available from the Inland Revenue series (albeit some time in arrears) and there are published indices of housebuilding construction costs. All these statistics have their limitations but they are indicative of trends and broad magnitudes and help to provide a conceptual framework for the analyst. Applying this theoretical model to specific companies may be more difficult, reflecting not just the managerial differences that complicate inter-company comparisons in any industry, but the different time structures that are chosen for the land bank. Thus, a long land bank housebuilder may be drawing on old low cost land at a time when a competitor is forced to draw on more recent high cost land. Conversely, the very short land bank company could be first out of any deep housing recession.

The possession of a long land bank, with a range of plot costs, gives its owner a greater degree of flexibility in taking profits. If rising selling prices are pushing profit margins above the long term level at which they can be sustained, the prudent management might increase the proportion of higher cost sites. When profitability comes under pressure, the cheaper sites can be brought into production.

The severity of this last recession has been such that nearly all housebuilders have made provisions against site value. Variations in treatment and the subjective element in land valuation (discussed in Chapter 8) should not be underestimated. As housebuilders emerge from recession, it may be difficult to establish the extent to which reported profits are a meaningful reflection of what is actually being achieved and what had been done to balance sheet values in earlier years. At the time of writing, eyes are turned to the recovery phase in the cycle. The land cost effect can, in certain conditions, lead to a recovery in trading profits (forget about exceptional write-offs) before there is any increase in either unit sales or house prices. As land prices have collapsed, so the replacement cost of land has been falling. On a constant volume of output, the old expensive land is gradually worked out and newer cheap land takes its place, thus restoring profits to 'normal' levels. Many housebuilders expected this phenomenon to occur during 1992 but were overtaken by yet more falls in volumes and selling prices.

If a working model of a housebuilder can be constructed to give at least a feel for the way profits react to well-identified movements in sales, prices and costs, no such model is practicable for the pure contracting operation. To be sure, there are indices of tender prices and costs to supplement the industry statistics on construction output and orders. However, they are probably less accurate than, say, house prices as an indicator and, with pre-interest margins which may be only 2 or 3%, small errors in the price or cost statistics will be greater than the total margin. Add to that the long term nature of the contract mix and the much greater variation in the profitability of individual contracts within one organisation, then the use of published tender price and cost data for short term modelling is a high risk exercise. For contractors, therefore, the guidance (official or otherwise) from within the company should not be ignored without strong reason. Despite the failure of most contractors to realise how severe the 1990–92 recession would prove, over a period of years it would be a brave analyst who pits his judgement against the Finance Director's knowledge – at least when the company is half way through the year for which the forecast is being made.

Smoothing the cycle

It may now be appropriate to return to the distinction drawn earlier between profits actually being earned, (which may or may not approximate to those produced by the internal management accounts), and those published in the audited accounts. It is still argued in the public arena that investors prefer the comfort of a smooth (and, of course, upwards) profit record to one which is cyclical; above all, nasty surprises are to be avoided. The inevitable response has been profit smoothing, usually built on conservative accounting in the good times to provide a cushion for both the inevitable lean years and to meet the unexpected. The techniques are not difficult: excessive provisions against work in progress on contracts which the management deems to be problematic, delay a claim receipt to just after the year end, and so on. Thus we have the paradox of companies in one of the most financially erratic industries producing long periods of unbroken profits growth.

One of the best examples of unbroken profits growth, uncomplicated by acquisitions, is Taylor Woodrow which had suffered a sharp setback to profits in 1960 resulting from problems on an overseas dam. From then onwards the profits movements were upwards only for a period of almost thirty years; generations of City analysts came and went without ever seeing a downturn in the company's profits. Within the component parts of the business, the expected cyclical fluctuations did occur with the overseas contracting profits in particular being feast and famine. A closer examination of the individual subsidiaries' accounts at Companies House during this period would show that whenever there was a sharp change in overseas profits, there was coincidentally a countervailing movement in the profits of the UK contracting company which could lead the outsider to deduce that a policy of smoothing was being practised. The most that could be observed from the pattern of profits declaration was that there were intermittent periods of fast growth and slow growth; the analyst was led, perhaps, to deduce that the first year of nominal increase in profits was a signal that Taylor Woodrow had temporarily gone 'ex-growth'. This psychology of profits declaration is far removed from any business school rationale of profits forecasting.

With the 30th record year of growth in prospect for 1990, we see that it was not to be. Along with the other contracting majors, profits collapsed and a loss was incurred in 1991. The interesting question is, were there signs available which would tell the outsider that whatever it was that had prevented profits falling in the previous 29 years was no longer going to work; if not, then any pretence at profit forecasting might as well be abandoned. Even without the benefit of hindsight (or perhaps just a little) there were signs, not necessarily of the extent of the fall that was about to happen, that the composition of Taylor Woodrow's profits flow was quite different than it had been even ten years previous.

The property activities had been progressively expanded since Taylor Woodrow Property was formed in 1964 and the rental flows actually increased the stability of group profits. However, during the 1980s, Taylor Woodrow became more active as a seller of property and in 1985 it changed its accounting policy to include profits on the sale of investment property, previously taken as an extraordinary item. By 1989, 40% of group profits came from the sale of investment properties; by adding development properties and private housing at the top of the market the figure rose to 66%. In contrast, contracting profits only accounted for 14% of the group total. The property tail was wagging the contracting dog, and even with the best will in the world, no finance director could create contracting provisions which would be large enough to act as a regulator of group profits. Anyone taking the view that property and housing were cyclical businesses would conclude that a fall in profits was likely. There are no great truths in this Taylor Woodrow digression, other than to emphasise the importance of a full understanding of the structure of the company for which the profits forecast is being made, and a reminder that what has always been does not necessarily continue.

On a more generalised view of profit taking, the pure contracting business will tend to bring in its profits at the tail end of the contract for all the reasons mentioned earlier in this book. It was pointed out, however, that falling turnover – although not immediately translated into falling contract profits – could cause a reduction in overhead recovery. There is an investor relations psychology which can reinforce that tendency: if a company thinks it has sufficient profit reserve to cushion it through an anticipated profit downturn then it will use that cushion to hold profits. Once management reaches the conclusion that the severity of the downturn is such that profits are bound to fall at some stage in the future, then the temptation is not to use the cushion and get the profits fall out of the way. If that change in expectations takes place after the end of the financial year, then the 'guidance' given to the City at what would be regarded as a late stage in the year, can prove surprisingly wide of the mark.

The same psychology is also observable in taking provisions against land holdings in the property development and housebuilding busin-esses. If the downturn in values is perceived to be short term in nature then there will be an incentive to 'sit it out'. However, once it is accepted that a provision against existing values is inevitable, the emphasis switches to making the provision immediately and even indulging in a degree of overkill to get all the bad news out of the way and prepare for a speedier profits recovery. It is a measure of the severity of this most recent downturn that many finance directors have had to indulge in overkill in two successive years.

Profits declarations coming out of a recession will also be subject to their own psychology. After the first 'artificial' bounce in profits (artificial in the sense that it will merely reflect the absence of provisions and other exceptional charges), recovery in published contracting profits will tend to lag behind the recovery in profits actually earned. Managements will be concerned to rebuild their inner financial reserves and it may be that the particular traumas associated with the 1990–92 recession will induce a financial over-conservatism that will colour the whole of the 1990s. In that event, the best lead indicator for a period of above average growth in profits might not be industry statistics but the appointment of a new (and young) finance director.

CONTRACTORS' CASH FLOW

Although this chapter has concentrated on the flow of profits, the generation of cash has always been an important aspect of the industry; indeed, some would argue that it is the only reason for being in the industry at all. It is therefore appropriate to conclude with a discussion of contractors' cash flow, its rationale, and whether the cash generation of the construction industry can be sustained indefinitely.

It is commonplace in the industry to state that contracting is cash positive, that it has negative capital employed, hence the argument that the cash flow could be used to fund capital intensive related industries – housebuilding, property development. SSAP25 required more detailed segmental analysis in accounts, calling in particular for analysis of assets by activity. As usual, there have been differing interpretations of the requirements, not only as to what constituted a separate activity but, additionally, at what level net assets should be calculated – equity, or including debt? Nevertheless, enough information is now available to see just how cash positive the contracting majors are.

Table 10.1 shows the contracting turnover and net assets of the nine largest quoted contractors, excluding those that are part of larger

Table 10.1 *Contracting turnover and assets of largest independent contractors*

	Turnover (£m)		Net assets (£m)	
	1991	*1992*	*1991*	*1992*
* Trafalgar House	2425	3234	(14.9)	21.4
† AMEC − Building & Civils	772	675	6.4	11.0
− M & E	1407	1362	89.2	83.9
† Laing	1422	1081	(111.2)	(108.0)
* Tarmac	1085	978	(72.7)	(130.4)
* Mowlem	1053	904	(69.9)	(45.0)
† Taylor Woodrow	1032	870	11.9	(58.1)
* Wimpey	790	693	(123.0)	(136.8)
* Costain	979	933	150.3	98.6
† Alfred McAlpine	395	344	25.2	22.2

* *Net assets includes net debt.*
† *Shareholders' funds.*

groups. There are, indeed, a significant number of negative or very small net asset figures. Both Trafalgar House and AMEC have large Mechanical and Electrical and Process Engineering operations which included fixed manufacturing assets. Interestingly, AMEC divides its construction business into mainstream contracting and M & E, highlighting the difference in capital utilisation between them. The other exception to the low capital base theory appears to be Costain but this reflects the substantial accumulated cash which has been left in the divisional accounts and lent on to fellow subsidiaries; the group accounts show inter-company loans approximately equal to the construction assets.

Finally, and at the risk of stating the obvious, a specialist contractor (i.e. with no other activities) cannot operate with negative capital employed. To say that contracting may not require positive capital on a day to day basis does not mean that clients or suppliers will deal with a contractor who displays no net worth. Positive capital has to exist, for all the obvious prudential reasons, and size of capital base will be an important consideration in obtaining work and in securing third party bonding. It is only in the contracting conglomerates, which present the totality of their group balance sheet to the outside world, that we can see how the surplus cash flow is used to support other activities.

Why should contractors be cash positive?

One simple explanation as to why contractors should be cash positive is by being paid before paying their own bills. Most overseas and many UK contracts have an advance or mobilisation payment and the contractor will also try to structure the contract so that the interim payments (normally made monthly on presentation of architects' or civil engineers' certificates) are weighted towards the earlier stages of the work flow. In contrast, some of the contractor's own payments will be made more than one month in arrears; builders' merchants could, for instance, be supplying two months' credit. Staff, of course, would be paid monthly and direct labour possibly more frequently. The grey area lies in the treatment of sub-contractors who may, in practice, be supplying much of the material and the specialist labour. Whatever the various forms of contract specify, the main contractor is normally in a stronger position than the sub-contractor and may use that strength to delay payments. In the March 1992 budget, the Government proposed a new clause in its contracts stipulating prompt payment of sub-contractors; at the time of writing, agreement had not been reached with the industry and the departments will draft their own clauses, although it is not thought that they will be binding. The pay-when-paid practice mentioned earlier (see page 161), whereby the main contractor will not pay the sub-contractor if it in turn is having funds withheld by the client, is necessary to protect the contractor's cash flow. Whether or not it is contractually legitimate, or 'unfair', is currently under intense scrutiny; test cases are going through the Courts.

For those contracts which proceed as planned, with no cost overruns, and payments on schedule, there seems no reason to expect the generation of positive cash flow to change, unless there are substantial alterations to standard contract payment terms or the length of credit extended by materials' producers and merchants. However, there is a tendency to assume that contracting is automatically cash positive, almost a divine right handed down by the 'contractor in the sky'. This assumption should be put in perspective. For a start, contractors are only temporarily cash positive. Eventually the bills have to be paid and the contractor is left with no more than his usual net profit. However, in a period of increasing turnover, the impact of the front end loading means that the cash flow will increase as new contracts are taken. When turnover falls, more of the workload will be at the finishing end, where there may even be a cash outflow, than at the mobilisation phase. Thus, although each contract remains cash positive in total, the cash balances accumulated earlier in the upswing of the contract cycle, will tend to dissipate. If those cash balances had been invested elsewhere, on the assumption that they were self-perpetuating, embarrassment follows. At this stage in the cycle, those contractors successful in increasing their market share are accused by their rivals of 'buying' work to protect their balance sheet. It is of little help to the outsider that success and failure both share the same lead indicator!

Simplistic numbers below illustrate the differing impact that front end loading of receipts can have on cash flow in the upswing and downswing of the cycle. Suppose there is a series of four year contracts earning 8% gross margins before overheads; the contract cash flow might be:

	Year 1	*Year 2*	*Year 3*	*Year 4*	*Total*
Contract costs	23	23	23	23	92
Cash receipts	36	25	20	19	100
Cash inflow – year	13	2	(3)	(4)	8
– cumulative	13	15	12	8	8

Now see how these cash flows will vary with a change in the number and distribution of contracts. Although each contract is assumed to be equally profitable, the time distribution produces a considerable variation in the cash flow. Indeed, it would not be difficult to devise a scenario in which the fall in orders produce a negative cash flow:

	Normal year		*Inflow of orders*		*Fall in orders*	
	No. of contracts	*Cash flow*	*No. of contracts*	*Cash flow*	*No. of contracts*	*Cash flow*
1st Year contracts	10	130	14	182	8	104
2nd Year contracts	10	20	12	24	10	20
3rd Year contracts	10	(30)	10	(30)	12	(36)
4th Year contracts	10	(40)	10	(40)	14	(56)
Total	40	80	46	136	44	32

The mathematical relationships also suggest that after a sharp inflow in orders, cash balances will eventually decline, even if turnover is maintained at these high levels. Conversely, at the bottom of the cycle, cash balances will rise if the inflow of new orders is maintained at the constant low level, once the old 'tail end' contracts have been worked through. All this, of course, assumes no change in margins during the course of the cycle.

In practice, the cash flow on individual contracts varies widely. Tender margins, i.e. the margin that the contractor expects to achieve on the basis of existing costs, will rise and fall with the levels of demand and supply, the same as for any other commodity. Achieved margins can sometimes run counter to the cycle if, for instance, buoyant markets lead to cost escalation or, in a recession, the contractor's buying power more than compensates for the deterioration in tender margin. But as we discussed earlier, individual contracts can also suffer delays and changes, whether through unforeseen conditions or variations requested – or insisted on – by the client. The excess costs incurred, even if eventually fully recouped from the client with the requisite profit margin, are rarely paid promptly. Indeed, 'discussion' may drag on for years. As it is not unknown for contract values to end up over twice the size of the original estimate, contracts with extensive claims can create short term cash deficits.

*T*he companies

CONTRACTORS

The analysis of any industry leads sooner or later to the compilation of lists of participants and the inevitable comparisons of size and performance. Unfortunately, the mix of business between companies and differing reporting practices precludes easy comparability. Some contractors will separately disclose one or more areas of activity: UK construction, overseas construction, housebuilding, property development, and mechanical engineering. Others will amalgamate one or more of these categories to the point where they might all be grouped as one class of business. Does one compare turnover, which in good times will understate the contribution from housebuilding, or profits, which may not exist in the depth of a recession? Table 11.1 shows the largest firms by size of contracting turnover, and for all its faults this table is probably a better guide than profits to relative weight in the industry. Despite the opening comments on differing definitions of contracting, the standard of segmental breakdown has improved considerably in recent years and most groups now separately disclose what most analysts would accept as

Table 11.1 *Contractors with turnover of £100 m in 1992 (contracting part of group only)*

		Turnover (£m)		
Company		1991	1992	
Q	Trafalgar House	2425	2448	Includes Trollope & Colls, Cementation Monk, Willett & Davy
Q	AMEC	2179	2037	Includes Fairclough, Press, Matthew Hall & James Scott
Q	Balfour Beatty	1830	1790	* Divisional total less estimated housing
Q	Bovis	1291	1400	* Divisional total less estimated housing
Q	Laing	1422	1081	
Q	Tarmac	1085	978	
Q	Costain	979	933	
Q	Mowlem	1053	904	
Q	Taylor Woodrow	1032	870	
Q	Wimpey	790	693	
P	Kier	614	545	
O	Hollandische Beton	c450	c425	Edmund Nuttall, Kyle Stewart, GA Group (1991)
Q	McAlpine, Alfred	395	344	
Q	Birse Group	312	343	
O	Norwest Holst	* 283	332	
Q	Tilbury Douglas	192	322	
P	Shepherd	387	280	
Q	Raine	264	259	Includes interior contracting
Q	Higgs and Hill	309	251	
P	Newarthill	421	224	Includes Robert McAlpine
P	Wiltshier	230	206	
P	Miller	183	195	
Q	Gleeson	199	183	Includes housing
P	Morrison	161	183	Includes small property development
P	Wates	200	170	Less estimated housing
P	Amey	145	165	
Q	Galliford	156	152	
P	Wilmot Dixon	158	152	
P	Biwater	160	150	
P	J. Murphy & Sons	137	140	
Q	Lovell	175	132	
P	Keller Group	113	132	
Q	Boot, Henry	130	129	Includes housing and property
P	Mansell	126	101	
P	Simons Construction	107	111	
Q	Try	100	103	

* *9 months only.*
Q — UK Quoted company or subsidiary thereof; P — Private; O — Subsidiary of overseas company.

mainstream contracting turnover as distinct from their housebuilding and property development activities. (The alternative approach is to include the totality of group turnover for any firm which falls within the contracting definition. This is the basis for the useful league tables produced by *Building* magazine each September which give the top 50 contractors by turnover, profits and employees.) Only a few firms still combine housebuilding with contracting and, for the two largest, estimates have been made of the contracting content.

Trafalgar House and AMEC, first and second respectively in the table, both have a heavy mechanical and electrical content and have moved further into process engineering than most of their competitors. They both test the distinction between the conventional building and civil engineer, and the mechanical engineer whose work is outside the remit of this book. Balfour Beatty benefits from power construction work which is peculiar to its parent, the BICC Group. And so one can go on explaining differences, but the list is intended to do no more than indicate who the leading firms are and their approximate standing to each other.

HOUSEBUILDERS

Table 11.1 excluded housebuilding to indicate the relative size of the construction businesses only. Nevertheless, housebuilding represents an important part of many contractors' portfolio of activities; in addition there are many specialist housebuilders who are not themselves contractors, but are still regarded as part of the contracting industry. To show their relative importance, a list of unit completions for the top 25 housebuilders has been extracted from the Credit Lyonnais Laing 1992 *Private Housebuilding Review*, updated with their co-operation. The list is shown in Table 11.2. The housebuilders are ordered according to completions in their 1992 financial years which, in the recession, favours those companies with an early year end. It also gives more weighting to those that build small rather than large houses, for each £m of turnover. For those wanting more corporate information on housebuilders, the Credit Lyonnais Laing review lists 75 firms with their financial records.

For the record, the constituents of the 'Contracting, Construction'

Table 11.2 *Private housebilders unit completions*

Housebuilder	Year end	1990	1991	1992	Parent
McLean/Tarmac	Dec	11 038	9327	7820	Tarmac
Wimpey	Dec	6 263	6380	5542	Wimpey
Beazer	June	5 372	5006	4948	Hanson
Barratt	June	5 950	4963	4706	Barratt Developments
Wilcon	Dec	2 350	2750	2880	Wilson Connolly
Lovell	Sept	2 823	3101	2571	Y.J. Lovell (Holdings)
Ideal	Sept	2 525	2258	2354	Trafalgar House
Persimmon	Dec	2 028	2324	2340	Persimmon
Bryant	May	1 600	1865	2330	Bryant Group
Westbury	Feb	2 266	2576	2277	Westbury
Laing	Dec	2 005	1670	2175	John Laing
Fairview	Dec	1 509	1952	1960	Hillsdown Holdings
Bellway	July	1 700	1518	1841	Bellway
Crest	Oct	1 367	1435	1465	Crest Nicholson
Hassall	June	1 354	1224	1429	Raine Industries
David Wilson	Dec	1 263	1324	1367	Wilson Bowden
Bloor	March	1 200	1300	1350	J.S. Bloor
Bovis	Dec	1 600	1481	1260	P & O
Alfred McAlpine	Oct	1 203	1189	1259	Alfred McAlpine
Redrow	June	957	1009	1105	Redrow Group
Tay	June	943	849	1094	Tay
Fairclough	Dec	1 395	1274	996	AMEC
McCarthy & Stone	Aug	1 002	937	938	McCarthy & Stone
Heron	March	765	766	900	Heron International
Berkeley	April	378	491	858	Berkeley Group

Source: Private Housebuilding August 1992, *Credit Lyonnais Laing, updated.*

sub-sector of the Financial Times-Actuaries (FT-A) All-Share Index follow. They have been marked as construction based (C); predominantly housebuilding (H); or plant hire (P), though the edges are occasionally a little blurred.

Name	
AMEC	C
Ashtead Group	P
Avonside	H
Barratt Developments	H
Bellway	H
Berkeley Group	H

Boot, Henry	C
Bryant Group	H
Costain Group	C
Countryside Properties	H
Crest Nicholson	H
Galliford	C
Gleeson, M.J.	C
Hewden-Stuart	P
Laing, John	C
McAlpine, (Alfred)	C
McCarthy & Stone	H
Maunders, John	H
Mowlem, John	C
Persimmon	H
Raine Industries	H
Tay Homes	H
Taylor Woodrow	C
Tilbury Douglas	C
Vibroplant	P
Westbury	H
Wilson Bowden	H
Wilson Connolly	H
Wimpey (George)	C

SHARE PRICE MOVEMENTS

The Contracting and Construction sub-sector of the FT-A All-Share Index has been the standard measure of performance in the sector since its commencement in 1964. The constituent companies include both the housebuilders and the contractors, whose fortunes may be moving in contrary directions, as well as the majors who straddle both activities. Commercial property development is thrown in for good measure. More detailed indices are available for those with access to the Datastream computer data banks, specifically a housebuilding index, and a separate contracting index excluding housebuilders.

A variety of relationships between the Stock Market movements and

175

supposedly causative economic and industry statistics can be postulated and this book takes just one as an example. The most commonly discussed relationship is between interest rates and the Construction Index. (To be more precise, the movements in the Construction Index are shown *relative* to the overall movement in the FT-A All-Share Index. This eliminates the effect of general movements in the Stock Market and isolates the specific preference or otherwise for construction shares; this is commonly referred to as the 'share price relative'.) Those who have worked for any length of time in the investment markets will be only too aware of how quickly day to day expectations of changes in interest rates can translate into share price movements in the building materials and construction sectors. Over a longer period of time, however, the relationship is not always clearly recognisable. Ultimately, share prices react to long term changes in profits, and expectations thereof. As has been indicated earlier, trends in industry activity and, hence, profits are often affected by interest rates, but not always, and with varying lags. Thus, there are periods of time when the interest rate/share price relationship works and times when it does not.

It can be instructive to look at specific time periods when the relative sector movements have been most marked and see the extent to which they mirror interest rate movements (see Fig. 11.1). Period 1 is the collapse of the early 1970s, when the underperformance of the construction sector began, around the time the banks' base rate began the

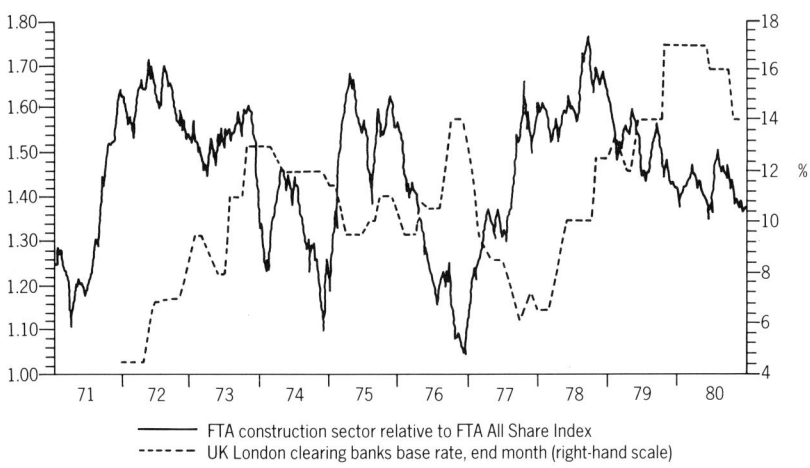

11.1 *Construction sector price index and interest rates 1970–80 (Source: Datastream).*

11.2 *Construction sector price index and interest rates 1980–93 (Source: Datastream).*

rise which was to take it from 5% in June 1972 to 13% in November 1973. So far so good. However, the downwards movement in share prices was only one third of the way through by the time that interest rates peaked and the sector continued to decline as the full extent of the profits falls and balance sheet deterioration became apparent; it was only towards the end of that downwards cycle of interest rates that a sharp recovery in the sector relative occurred, almost taking the sector relative back to its previous peak. However, the 'reality' eventually made itself felt and 1976 ended with the sector relative lower than at any point in that recession.

Period 2 (shown in Fig. 11.2) starts with the mid-1980s rise in interest rates which began early in 1984 and lasted for a couple of years. On that occasion, the sector relative peaked a good year in advance of the turn in interest rates (anticipation?). For the next few years the relationship appears to work as one might expect. The 1985 peak in rates marks the start of the recovery in the sector price relative which continues until interest rates bottom out in mid-1988. Although profits were to continue rising strongly for at least another year, investors could see the implications of the change in interest rate policy and almost simultaneously the share price relative began what was to be a precipitous decline. Once interest rates peaked, and then as they were successively reduced during 1991 and 1992, there were corresponding mini-rallies in the sector but, despite what amounted to a halving of interest rates, the main thrust of construction share price relatives continued to be inexorably downwards.

It can be seen, therefore, that the response of the construction sector price relative to interest rate movements can be reasonably close for some cycles though not for others. Where the interest rate indicator particularly appears to fail is in indicating the recovery point in those cycles which have proved exceptionally deep – the early 1970s and this most recent downturn – when the reality of the recession far outweighs the anticipation of the recovery. The explanation is simple enough: if the Government is determined to check an economic upturn, it can keep raising interest rates until it succeeds. Unfortunately, the reverse is not always true; you can lead the economic horse to the pool of lower interest rates, but the horse will occasionally have other things on its mind.

The discussion of interest rates and the construction sector price movements is no more than one example of the relationships that can be plotted. It was chosen because it is the most frequently cited, and to demonstrate that what might seem to work in the short term (perhaps within the Stock Exchange Account) does not necessarily hold good in the long term. It works in some cycles but not in others: the interest rate enthusiast investing in the contracting sector at the end of 1990, as rates began to fall, would have lost half his money over the succeeding two years, and still have been right on interest rates.

Conclusion

If it was so, it might be; and if it were so, it would be: but as it isn't, it ain't. That's logic. (Through the Looking-Glass)

Looking back through the chapters of this book, it is clear to the author that the text has, from time to time, taken on a life of its own, as particular interests and preoccupations have been followed while others, perhaps, have been skimped or even neglected. This is an observation rather than an apology, for economic and investment analysis remains an art as much as a science; the practising analyst knows how important subjective judgement is in the exercise of his trade, so the author has some excuse if he has exercised subjectivity in determining the content of this book.

Most analysts think that their chosen industry specialisation is more interesting than the rest of the market; indeed, analysts can indentify so closely with their subject that they can become blinkered and defensive. (Just experiment by criticising bank lending practices to a banking analyst or exorbitant water rate increases to a utilities analyst.) The construction industry does, of course, have a special attraction. Economically, the analyst must become involved in matters of public policy as it determines public spending; in corporate investment decisions; in the utilities; and, of course, in a whole range of consumer spending from the biggest of all, the house, right through to minor repair and maintenance expenditure.

Because of its pivotal position in the economy, and the social importance of one of its most long-standing products – the house – the construction industry is extensively, even interminably, analysed. But more than anything, it is the weight of statistical data available to the researcher that can be an open encouragement to excessive analysis. Nowhere is this more apparent than in the housing market where stock, transactions, new construction, finance, prices, and costs can all be analysed every which way. Contrast this with repair and maintenance, which accounts for some 41% of recorded construction output, and where private housing repair and maintenance exceeds the value of new private housing construction output. Only the most limited statistical data exists and the independent analysis on this area of the construction market is, accordingly, minimal. In practice, many forecasts of total construction output will be highly detailed for 20% of the content, less detailed for a further 40%, and sketchy for the balance.

The proposition was, therefore, that those statistics which were available should be analysed. To facilitate this, the chapters on statistical sources (Chapter 4) and forecasting (Chapter 9) seek to guide the user through those statistics which are most frequently quoted, and which are of most help in making a forecast (not necessarily the same thing). A constant theme has been that users should not take these statistics, their basic building blocks, as gospel. The nature of their compilation, particularly those statistics that are price and value based, does not make for perfect accuracy. The instance was quoted of the constant price construction output series which showed total construction output still rising through 1990, a year in which virtually every supporting statistic was in strong decline. For a more continuing example, it is known by all that a large proportion of repair and maintenance expenditure is unrecorded, introducing who knows what errors into the total construction output figures.

Do these statistical weaknesses matter? Insofar as the purpose of the forecast is to identify trends, then probably not. However, it is to be hoped that readers of this book will, as well as learning to find their way around the industry, also develop a healthy regard for the limitations of their raw material and to nurture their subjective as well as objective judgement.

When one moves from the industry to the corporate sector, the uncertainties and qualifications multiply. The accounting within the construction industry, the supposedly definitive measure of the health of the corporate body, is peculiarly subjective. Much rests on the year end valuation of work in progress with decisions on the extent to which directors can 'reasonably foresee' the outcome of a contract, the need for provisions against contracts which may or may not be expected to be loss making, and the treatment of claims which, in civil engineering, can exceed the original value of the contract. Even the housebuilders taking their profit as each house is sold also face subjective decisions on the valuation of site work in progress and the allocation of costs between plots or phases of the site. For contractors and housebuilders alike, changes in assumptions or expectations part way through a contract or site can have a significant effect on the profits of the year in which the assumptions are changed. All this is before the subjective decisions which have been made during 1990–92 on the correct carrying values for development assets. And, as the recovery develops, companies are not required to disclose the extent to which earlier provisions against current assets have been credited back to profits.

That, then, is the basis on which rests that profits numbers that the analyst is trying to forecast. It is assumed that the reader has read more general works on profits forecasting and the forecasting chapter in this book tries to provide a construction framework. In truth, the practice of profits forecasting is far removed from the theory that its practitioners sometimes like to present. Companies usually encompass a number of different activities; accounting information of the most basic kind (e.g. segmental information) is not presented on a basis which allows for proper comparison between companies; and there is rarely access to the intermediate data that would permit sensible model building.

Formal and informal contact and liaison between the forecaster and the company became widespread during the 1970s and intensified during the 1980s. The relationships have become so close that 'The Company says . . .' has threatened to replace 'I think . . .', despite the fact that few companies knew what was happening to themselves through 1990 to 1992. Attitudes may now be changing as regulatory pressures call into question the relationships between company and analyst. In any event, the author is not convinced that the 'guidance' emanating from the companies on the size of the profits upturn will be as forthcoming or reliable as in previous cyclical upturns. To the extent that the analyst will therefore need to prepare his ground more thoroughly, it is hoped that this book will provide, if we can extend the construction metaphor, at least a part of that foundation.

*B*ibliography

The references in this bibliography are grouped under broad subject headings although some material would readily fit into more than one category.

HISTORY AND STRUCTURE OF THE CONSTRUCTION INDUSTRY

Chrimes, M.: *Civil engineering 1839–1889: A photographic history*, Stroud, 1991.

Colvin, H.M.: *A biographical dictionary of English architects 1660–1840*, 1954.

Cooney, E.W.: 'The origins of the Victorian master builders', *Economic History Review*, **8**, 1959, pp. 167–76.

Helps, Sir A.: *Life and labours of Brassey*, London, 1872 (reprint 1969).

Hobhouse, H.: *Thomas Cubitt master builder*, London, 1971.

Merrett, S.: *Owner occupation in Britain*, London, 1982.

Middlemass, R.K.: *The master builders*, London, 1963.

Norrie, C.M.: *Bridging the years – A short history of British civil engineering*, London, 1956.

Port, M.H.: 'The Office of Works and Building Contracts in Early C19 England', *Economic History Review*, 1967, pp. 104–109.

Postgate, R.W.: *The builders' history*, London, 1923.

Powell, C.G.: *An economic history of the British building industry 1815–1979*, London, 1980.

Smyth, H.: *Property compaines and the construction industry in Britain*, London, 1985.

Spender, J.A.: *Weetman Pearson First Viscount Cowdray 1856–1927*, London, 1930.

Whitehouse, B.: *Partners in property*, London, 1964.

Summerson, J.: *The London building world of the 1860's*, London, 1973.

STATISTICAL SOURCES

Butler, A.D.: *'New price indices for construction output statistics'*, *Economic Trends*, no. 297, July 1978.

Confederation of British Industry: *Industrial Trends Survey*, quarterly.

ECONOMIC TRENDS

Fleming, M.C.: *Reviews of United Kingdom statistical sources vol. XII: Construction and the related professions*, Oxford, 1980.

Fleming, M.C.: *Spon's guide to housing, construction and property market statistics*, London, 1986.

Government Statistical Service: *Housing and construction statistics*, annual and quarterly.

Heggs, P. and Holmans, A.: 'Number of property transactions in England and Wales', *Economic Trends*, no. 452, June 1991.

Housebuilders' Federation: *Housing Market Report*.

Jones Lang Wootton: *Central London Office Survey*.

King, Sturge & Co.: *Industrial Floorspace Survey*.

National House-Building Council: *Private House-building Statistics (Quarterly)*.

HOUSE PRICES

Department of the Environment: 'A new index of average house prices', *Economic Trends*, no. 348, October 1982, pp. 134–8.

Fleming, M.C. and Nellis, J.G.: 'House-price statistics for the United Kingdom: a survey and critical review of recent developments', *Environment and Planning A*, vol. 17, 1985, pp. 297–318.

Fleming, M.C. and Nellis, J.G.: *Spon's house price data book*, 1987.

Halifax Building Society: *House Price Bulletin*.

Housing Finance: The Quarterly Economics Journal of the Council of Mortgage Lenders (the successor body to the Building Societies' Association). Quarterly.

Nationwide Building Society: *Quarterly House Price Bulletin*.

LONG TERM HOUSING

Corner, I.E.: *Household demography and the effective demand for new housing*, Building Research Establishment, February 1991.

Corner, I.E.: *Modelling housing demand*, Building Research Establishment, September 1991.

Department of the Environment: *English House Condition Survey, 1991*, 1993.

Department of the Environment: *Household Projections England 1989–2011*, 1991.

Department of the Environment: *Housing Policy*, 1977.

Government Statistical Service: *1989-based National Population Projections Series PP2 No. 17*, 1991.

Government Statistical Service: *Population Trends*.

Holmans, A.: 'A forecast of effective demand for housing in Great Britain in the 1970's', *Social Trends*, no. 1, 1970.

Holmans, A.: 'The 1977 Housing Policy Review in Retrospect', *Housing Statistics*, vol. 6, July 1991.

Holmans, A. and Nandy, S.: 'Household formation and dissolution and housing tenure: A longitudinal problem', *Social Trends*, no. 17, 1987. NB: This article examines the results of a detailed sample study on household movements between the 1971 and 1981 censuses.

King, D.: *The 'Demographic Bulldozer' – myth or reality*, Anglia Polytechnic, April 1991.

CONTRACTS AND ACCOUNTING

Capon, G.: *An industry accounting and auditing guide to the construction industry*, Institute of Chartered Accountants, London, 1990.

Holmes, G. and Sugden, A.: *Interpreting company reports and accounts* (4th edn.), 1990.

May, Sir A.: *Keating on building contracts* (5th edn.), London, 1991.

Tillett, D. and Hunt, D.: *An industry accounting and auditing guide property company accounts*, Institute of Chartered Accountants, London, 1991.

FORECASTING

Lewis, P.: *Building cycles and Britain's growth*, London, 1965.

Richardson, H. and Aldcroft, D.: *Building in the British economy between the Wars*, London, 1968.

(Although both of these books may, at first sight, be of primary interest to the economic historian, they both contain excellent academic analyses of the economic relationships within the construction industry.)

Adams, A.: *Investment*, 1989.

Breedon, F.J. and Joyce, M.A.S.: *'House prices, arrears and possessions'*, Bank of England Quarterly, May 1992, pp173–9.

Copeland, T., Koller, T. and Murrin, J.: *Valuation: measuring and managing the value of companies*, New York, 1990.

Cottle, S., Murray, R.F. and Block, F.E.: *Graham and Dodd's Security Analysis* (5th edn.), New York, 1988.

Fridson, M.S.: *Financial statement analysis, a practitioner's guide*, New York, 1991.

Merrett, S.: *Owner occupation in Britain*, 1982. (Analysis of the housing market, with particular reference to post-war developments.)

Public Expenditure Analysis to 1995–96: Statistical Supplement to the Autumn Statement, Cmd. 2219, January 1993.

Winfield, R.G. and Curry, S.J.: *Success in investment* (4th edn.), 1991.

COMPANIES

Building, September 1992.

Guide to Financial Times Statistics Financial Times Business Information, (3rd edn.), London, 1991.

New Civil Engineer, Contractors File Supplement, 1992.

Credit Lyonnais Laing Private Housebuilding Annual, London, August 1992.

*I*ndex